Jinjie Chen

Local Investigation of Single Magnetic Molecules with Scanning Tunneling Microscopy

Experimental Condensed Matter Physics
Band 21

Eine Übersicht aller bisher in dieser Schriftenreihe erschienenen Bände finden Sie am Ende des Buchs.

Local Investigation of Single Magnetic Molecules with Scanning Tunneling Microscopy

by
Jinjie Chen

KIT Scientific Publishing

Dissertation, Karlsruher Institut für Technologie
KIT-Fakultät für Physik

Tag der mündlichen Prüfung: 09. Februar 2018
Referenten: Prof. Dr. Wulf Wulfhekel, Prof. Dr. Wolfgang Wenzel

Impressum

KIT Scientific Publishing

Karlsruher Institut für Technologie (KIT)
KIT Scientific Publishing
Straße am Forum 2
D-76131 Karlsruhe

KIT Scientific Publishing is a registered trademark
of Karlsruhe Institute of Technology.
Reprint using the book cover is not allowed.

www.ksp.kit.edu

Print on Demand 2018 – Gedruckt auf FSC-zertifiziertem Papier

ISSN 2191-9925
ISBN 978-3-7315-0819-9
DOI 10.5445/KSP/1000084345

Local Investigation of Single Magnetic Molecules with Scanning Tunneling Microscopy

Zur Erlangung des akademischen Grades eines
Doktor der Naturwissenschaften
von der Fakultät für Physik
des Karlsruher Institut für Technologie (KIT)

genehmigte
Dissertation

von
M.Sc. Jinjie Chen
aus
Changzhou, Jiangsu (China)

Tag der mündlichen Prüfung:	09. 02. 2018
Referent:	Prof. Dr. Wulf Wulfhekel
Korreferent:	Prof. Dr. Wolfgang Wenzel

Contents

1 Introduction

Novel materials are believed to be potential triggers for the next industrial revolution. In contrast to the roles permanent magnets played for electricity in the second industrial revolution and magnetic storage played for information technology in the third industrial revolution, without a doubt magnetic materials will again demonstrate their significances to spintronics in the next industrial revolution. With respect to the desires of smaller size, higher capacity and lower energy consumption, integrated circuits have been improved constantly following Moore's law for the last 60 years, which was proposed in 1965 to conclude and predict that the capability of integrated circuits doubles every 18 months. In 2017, IBM, in partnership with Samsung and GlobalFoundries, announced that the processes to build chips with 5 nm structure pitch (i.e. the size of a single transistor) had been developed. It means that we are marching towards a new era of fabrications on the nanometre scale. Accordingly, it is well believed that innovations in magnetic materials will emerge.

Two strategical approaches have been established to manufacture nano-structures. In the "top-down" approach, homogeneous materials are shaped to desired dimensions with micro-patterning techniques, e.g. photolithography. This approach is widely used in modern silicon-based chip fabrication. However, the limited physical dimensions cause tunneling currents and other technical problems as the size of transistors approaches nano-scale. Hence, the second approach called "bottom-up" is drawing more and more attention, since the problems described above can be avoided by engineering nano-structures from assemblies of single units. This "bottom-up" method has the advantage to produce devices in parallel and lower cost in comparison to the "top-down" method. Obviously, the density of information storage can also be much higher if each bit is stored in a single unit.

Single magnetic atoms and molecules are attracting much interest, since they are naturally considered as the smallest physical unit to be assembled in e.g. spintronic devices, data storage devices and quantum computers. In contrast to atoms in free space, the ions in magnetic molecules are influenced by the electrostatic interaction with the organic ligands. This opens up means to manipulate the characteristics of the magnetic molecules by modifying the ligands. Together with the spin-orbit interaction, the crystal field contributed by the ligands leads to magnetic anisotropy, or more precisely to a splitting of the magnetic multiplet of the $2S+1$ states even in absence of a magnetic field [1, 2, 3]. This anisotropy is crucial to achieve a stable magnetization. Additionally, the advantages of organic materials, such as low cost, light weight, high mechanical flexibility, sufficient resources and abundant chemical properties are highly valued. Moreover, almost all of the fundamental physical effects taking place in solids have also been observed in organic systems. With this, organic materials are potential candidates to replace present silicon-based materials. Consequently, magnetic molecules will play an incomparable role in the field of organic spintronics.

Among single magnetic molecules, complexes containing transition metal ions are most significant, as almost all of the magnetic materials in nature involves transition metal elements. Additionally, thanks to the various chemical valences and more than one metastable states, transition metal ions usually present tunable properties. In contrast to the molecules containing multiple transition metal ions, the ones containing only single ions attract special interest in research, because they are not only easier to synthesize but also serve as prototypes to understand the physics behind single magnet molecules. These molecules are referred to as single-ion magnets in literature, although they are not real magnets yet.

Due to nano-scale applications, experimentalists are motivated to investigate single magnetic molecules individually, which can be achieved with break junctions or scanning tunneling microscopy (STM). Especially in an STM, the functions of the molecules can be traced back to the different components within a molecule due to the high spatial resolution of STM. The spin states and magnetic properties of molecules can be revealed by scanning tunneling spectroscopy (STS). Towards functional devices, molecules must be placed on substrates, which not only support but also interact with the molecules on top. For instance, the Kondo effect may emerge when magnetic molecules sit on metallic surfaces [4]. Exchange coupling arises when both of them are magnetic. Essentially, these kinds of interactions in turn broaden the potential applications for these systems.

In this dissertation, an introduction to related theoretical backgrounds and employed experimental techniques are given at first in Chapters 2 and 3, respectively. Subsequently, the findings on various molecules are presented in Chapters 4, 5 and 6. At each beginning of these three chapters, a brief review of the corresponding topics and an abstract of our work are given before presenting the details. Finally, a conclusion and a prospect is expressed. The main discoveries of this dissertation are listed in the following.

In Chapter 4, molecules containing single 3*d* ions are studied with STM. In the first project, we systematically studied the zero-field splitting (ZFS) and the Kondo effect in mechanically flexible M-bis(2,2,6,6-tetramethyl-3, 5-heptanedionato) (M = Co, Ni and Cu) molecules (for short $M(thd)_2$) on $Cu_2N/Cu(100)$ surface. Interestingly, we find that the ZFS can be abruptly switched by the sudden formation or breaking of a molecular contact. Consequently, the energetic order of spin states, namely the magnetic anisotropy can be completely changed. This implies that this system can potentially be used to prepare superposition states, if the coherence time is long enough. To the best of our knowledge, it is the first time to contact magnetic molecules with an STM and measure transport in magnetic fields at temperatures as low as 30 mK. In the second project, the adsorption geometries of manganese phthalocyanine (MnPc) molecules deposited on Cu(100) and Co/Cu(100) are analysed. Moreover, the magnetic properties of MnPc are revealed by the observations of a Kondo resonance of MnPc on Cu(100) and spin-flip inelastic excitations of MnPc on Co/Cu(100).

In Chapter 5, as examples of single molecules containing 4*f* ions, a mono-nuclear lanthanide complex $Tb(thd)_3$ and a binuclear lanthanide complex $[Tb(tmhd)_3]_2bpm$ were transferred onto various metal substrates without decomposition or contaminations. The STM topographies indicate that these complexes form assemblies with distinct reconstructions on different substrates.

In Chapter 6, in order to investigate the exchange coupling between spin crossover molecules and ferromagnetic layers, $Fe(1,10\text{-phenanthroline})_2(NCS)_2$ (Fe-phen for short) molecules were deposited onto $Fe_2N/Cu(100)$ and lanthanum strontium manganite (LSMO) films. On the former system, Fe-phen molecules were reversibly switched between a high-spin and a low-spin state by tunneling currents at 5 K. On the latter one, the magnetization of the hybrid system was studied in a temperature range from −120 °C to room temperature with the longitudinal magneto-optic Kerr effect. Unfortunately, the results of this brief experiment do not provide any evidence for a potential coupling between Fe-phen molecules and the ferromagnetic LSMO substrate.

2 Theoretical background

Condensed matter physics is a subject that has a wonderful cooperation between theory and experiment. On one hand, many novel experimental discoveries have been perfectly interpreted by theoretical models. On the other hand, many courageous theoretical predictions have been confirmed by tricky experimental designs. As an experimentalist in this field, before carrying out the experimental project, necessary related knowledge must be established to design an experiment and analyse the results. Frequently, the experiment however does not progress as planned. In that case, an abundant background knowledge and a rich experience help to understand the complications and to modify the plan accordingly. Actually, this is how numerous important phenomena were discovered. In this chapter, the theoretical concepts related to this dissertation will be introduced in order to strengthen the motivation and to be able to analyse the result better.

2.1 Ligand field theory

Single magnetic molecules that attract our interest contain a central metal ion surrounded by organic ligands. Since the the central ion and electrons on it are influenced by the electrostatic interaction with the surrounding ligands, the degenerate orbitals of the ion split. The energetic order of the new orbitals and the magnitude of the splitting strongly depend on the symmetry of the system, the identity of the ligand and the charge on the central ion. To represent the non-spherical "field" resulting from the ligands, the ligand field theory (LFT) combining crystal field theory and molecular orbital theory was developed [5]. The successful applications prove that LFT is a powerful tool to describe transition metal coordination complexes, especially the process of chemical bonding. As this dissertation focuses on the magnetic properties of molecules, our attention is the states of electrons on the central ion rather than the chemical bonds in the whole molecule. Therefore, only crystal field theory (CFT) without molecular orbital theory is a sufficient model to understand the static electric field contributed by surrounding charges, thence to deliver the splitting of the degenerate electron orbitals on the central ion.

A qualitative understanding about the crystal field breaking the degeneracy of electron orbitals can be expressed as follows. Since the orbitals of the central transition metal ion are distributed in space differently, some of them are closer to the surrounding charges on average, while the others are further away. Thus, the electrons on different orbitals feel different strength of electric field and experience different interaction with surrounding charges. For instance, the electron on an orbital close to a negatively charged ligand feels a repulsion, which explains why the energies of this orbital is raised. In comparison, the energies of the orbitals further away from ligands are less increased. As an additional term $\mathscr{H}_{\mathrm{CF}}$ contributed by crystal field should be added to the Hamiltonian, the original eigenstates of an electron in a spherical field are mixed. Fortunately, the crystal field is usually not sufficient to mix orbitals of different subshells, which means the orbital quantum number l is till a good quantum number, but the magnetic quantum number m is not. Since inner shells are screened by outer ones, crystal field only affects outer shells. Moreover, the s and p orbitals are frequently not localized on the ions and have high symmetry, thus CFT is frequently applied on d or f orbitals, although the effect on the latter one is already relatively small because of the screening.

To describe the crystal field quantitatively, one should take all of the involved charges into account. Their contributions to the static electric potential $\psi_{\mathrm{CF}}(\boldsymbol{r})$ around the central ion can be integrated as following:

$$\psi_{\mathrm{CF}}(\boldsymbol{r}) = \int \frac{\rho(\boldsymbol{r}')}{4\pi\varepsilon_0|\boldsymbol{r}'-\boldsymbol{r}|} d^3\boldsymbol{r}'. \tag{2.1}$$

Here, a spherical coordinate system centred on the ion is used, thus $\boldsymbol{r} = (r, \theta, \phi)$. $\frac{1}{|\boldsymbol{r}'-\boldsymbol{r}|}$ can be expanded in spherical harmonics:

$$\frac{1}{|\boldsymbol{r}'-\boldsymbol{r}|} = \sum_{n=0}^{\infty} \frac{4\pi}{2n+1} \frac{r^n}{r'^{n+1}} \sum_{m=-n}^{n} (-1)^m \boldsymbol{Y}_n^{-m}(\theta', \phi') \boldsymbol{Y}_n^m(\theta, \phi). \tag{2.2}$$

Hence, the electric potential of crystal field can be expressed as:

$$\psi_{\mathrm{CF}}(\boldsymbol{r}) = \frac{1}{\varepsilon_0} \sum_{n=0}^{\infty} \frac{r^n}{2n+1} \sum_{m=-n}^{n} \boldsymbol{Y}_n^m(\theta, \phi) \gamma_{nm}, \tag{2.3}$$

where

$$\gamma_{nm} = \int \frac{\rho(\boldsymbol{r}')}{r'^{n+1}} (-1)^m \mathbf{Y}_n^{-m}(\theta', \phi') \, d^3 r'. \tag{2.4}$$

Therefore, the Hamiltonian contributed by crystal field can be represented by:

$$\mathscr{H}_{\mathrm{CF}} = \int \rho_{ion}(\boldsymbol{r}) \psi_{\mathrm{CF}}(\boldsymbol{r}) \, d^3 r, \tag{2.5}$$

where $\rho_{ion}(\boldsymbol{r})$ refers the electron charge distribution of the central ion under concern.

To solve this Hamiltonian, the exact charge distribution $\rho(\boldsymbol{r}')$ in the environment should be known. In practice, the charge distribution is usually approximated by a point charge. Moreover, some terms in Equation 2.3 vanish for particular symmetric charge distributions. Without doubt, the crystal field strongly depends on the symmetry of the solid or the geometry of the molecule. So the CFT is always discussed for a specific symmetric structure. For instance, it is frequently the case that the considered ion is located in an octahedral or a tetrahedral cage in a solid, which is decorated by six or four anions on the vertexes. In the following, the *d* orbitals of an ion sitting in the centre of an octahedral cage will be discussed as an application of CFT.

2.1.1 Crystal field theory for octahedral cages

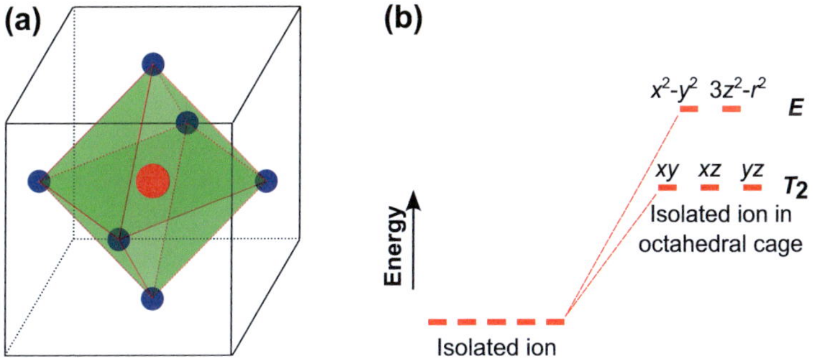

Figure 2.1: (a) Schema of an octahedral cage (green) with an central ion (red dot) surrunded by six anions (blue dots). (b) In presence of the corresponding crystal field, the five *d* orbitals split into two subgroups named as E and T_2.

As sketched in Figure 2.1 (a), a prototypic structure to apply CFT is a cubic crystal system where a metal ion (red dot) is centred in octahedron with six anions (blue dots) as the vertexes. From the point of view of group theory, the five-dimensional representation Γ_d of octahedral operations of d orbitals reduces to two irreducible representations $\boldsymbol{E}$ and $\boldsymbol{T_2}$ [6], which are two- and three-dimensional, respectively:

$$\Gamma_d = \boldsymbol{E} \oplus \boldsymbol{T_2}. \tag{2.6}$$

For the Hamiltonian of an isolated ion, the basis can be chosen in many ways. One basis in the form of $\boldsymbol{Y}_l^m \pm \boldsymbol{Y}_l^{-m}$ consists exclusively of real-valued:

$$\left.\begin{array}{rcl} r^2[\mathbf{Y}_2^2(\theta,\phi)+\mathbf{Y}_2^{-2}](\theta,\phi) & \sim & x^2-y^2 \\ r^2\mathbf{Y}_2^0(\theta,\phi) & \sim & 3z^2-r2 \end{array}\right\} \quad \boldsymbol{E} \tag{2.7}$$

$$\left.\begin{array}{rcl} r^2[\mathbf{Y}_2^2(\theta,\phi)-\mathbf{Y}_2^{-2}](\theta,\phi) & \sim & xy \\ r^2[\mathbf{Y}_2^1(\theta,\phi)-\mathbf{Y}_2^{-1}](\theta,\phi) & \sim & xz \\ r^2[\mathbf{Y}_2^1(\theta,\phi)+\mathbf{Y}_2^{-1}](\theta,\phi) & \sim & yz \end{array}\right\} \quad \boldsymbol{T_2}. \tag{2.8}$$

Actually, the top two and bottom three real functions form the basis of the $\boldsymbol{E}$ and $\boldsymbol{T_2}$ representation, respectively. The angular distribution of these five wave functions is presented in Figure 2.2. Note that the orbital angular momenta of these five orbitals are quenched, which means $\langle d_{orbit}|\,\boldsymbol{L}\,|d_{orbit}\rangle = 0$, where $\boldsymbol{L}$ is the operator of orbital angular momentum, d_{orbit} can be any orbital presented in Equation 2.7 and 2.8.

Group theory is a powerful tool to predict the splitting of energy levels and degeneracy of the new levels resulting from the crystal field of a particular symmetry. However, it can tell us neither the size of the splitting Δ, nor the energetic order of new levels. To know those precisely, we have to find eigenfunctions of the Hamiltonian described by Equation 2.5. Taking the octahedral cage as an example again, if the anions on the vertexes are approximated as points with effective charges of $-q$, by omitting the pre-factor, the crystal field Hamiltonian can be written as:

$$\mathscr{H}_{\text{CF\_oct}} = \sum_{i=1}^{6} \frac{-q}{|\boldsymbol{r}-\boldsymbol{R}_j|}, \tag{2.9}$$

where $\boldsymbol{R}_j$ is the location of each anion.

Qualitatively, unlike in a spherical field of an isolated ion, electrons in the octahedral cage experience additional repulsion from surrounding anions. Thus the energies of all d orbitals are increased. Furthermore, the d orbitals lose their degeneracy because of the different strength of repulsion. As presented in Figure 2.2, $x^2 - y^2$ and $3z^2 - r^2$ orbitals of the $\boldsymbol{E}$ representation directly point to six anions located on the axes. In contrast, xy, xz and yz orbitals of the $\boldsymbol{T_2}$ representation are further away from the anions since they are distributed between axes. Consequently, the $\boldsymbol{E}$ orbitals show higher energy than $\boldsymbol{T_2}$ orbitals. The energy levels are sketched in Figure 2.1 (b) as a comparison between an isolated ion and an ion located in octahedral cage.

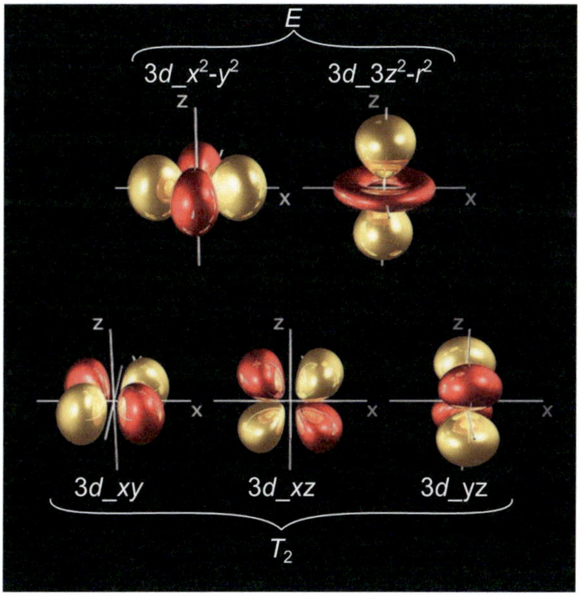

Figure 2.2: The shape of the five real wave functions regarded as d orbitals, which compose the basis of the $\boldsymbol{E}$ and $\boldsymbol{T_2}$ representation in an octahedral symmetry. This figure is originally sketched by Dr. Mark J. Winter.

2.2 Zero-field splitting

As discussed before, some particular orbitals have relatively lower energy in comparison to the others in the presence of crystal field. If the system displays strong spin-orbit coupling (SOC) $\Lambda \boldsymbol{L} \cdot \boldsymbol{S}$ at the same time, particular spin states are preferred in order to minimize the energy. In other words, the degeneracy of $2S+1$ spin micro-states breaks in a system experiencing crystal field and SOC. Since the spin states split even in the absence of magnetic field, this phenomenon is called zero-field splitting (ZFS) [2]. In comparison,

Zeeman splitting refers to the separation of spin states according to the values of spin quantum number (m) in the presence of a magnetic field. Obviously, the ZFS affects the magnetic properties dramatically by tuning the configuration and energetic order of spin states. Usually, ZFS occurs to the unpaired electron on transition metal ions, particularly in mono-nuclear complexes. Without doubt, the knowledge about ZFS is essential for investigating and interpreting the properties of single magnetic molecules.

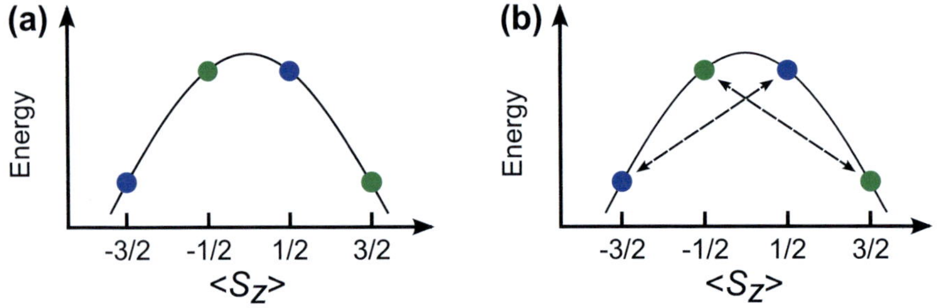

Figure 2.3: Schematic configuration of the spin states of an ion with total spin $S = 3/2$ under a phenomenological Hamiltonian with only a D term (a) and an E term added (b). The downward parabola corresponds to a negative D parameter. The dashed arrows indicate the states mixing ascribed to the E term.

Phenomenologically, ZFS can be described by model Hamiltonians in the form of a sum of effective spin Hamiltonians. The spin Hamiltonians directly reveal the splitting of the magnetic energy levels resulting from the effects of crystal field and SOC, when orbital angular momenta are quenched in the presence of crystal field. The magnetic property of the ion strongly depends on how the spin states split and how the energy levels of $2S+1$ micro-states distribute, which is determined by the spin Hamiltonians. One simple and widely used phenomenological Hamiltonian can be expressed as:

$$\begin{aligned}\mathscr{H}_{\text{eff}} &= D\boldsymbol{S}_z^2 + E(\boldsymbol{S}_x^2 - \boldsymbol{S}_y^2) \\ &= D\boldsymbol{S}_z^2 + \frac{E}{2}(\boldsymbol{S}_+^2 + \boldsymbol{S}_-^2),\end{aligned} \tag{2.10}$$

where D and E are parameters related to anisotropy, $\boldsymbol{S}_x$, $\boldsymbol{S}_y$, $\boldsymbol{S}_z$, $\boldsymbol{S}_+$ and $\boldsymbol{S}_-$ are the spin operators of x-, y-, z-component, raising and lowering, respectively. The first term describes the axial anisotropy, while the second term offers information about transversal anisotropy in the xy plane. In order to minimize the energy, larger S_z is preferred for ground state when D is negative. Then, the ion presents an out of plane easy axis. In contrast, a positive D corresponds to an in-plane easy axis. To understand Equation 2.10, we can take the case of a total spin $S = 3/2$ as an example. Four eigenstates, labelled $S_z = -3/2$, $-1/2$, $1/2$, and $3/2$, of the spin of an isolated ion are mixed and redistributed on energy levels under this Hamiltonian. As depicted in Figure 2.3 (a), the four states are split into two doublets. When the E term is taken into account, it mixes the states with $\Delta S_z = 2$ (see Figure 2.3 (b)). Therefore the spin quantum number m is not a good quantum number any more; new eigenstates are linear combinations of $S_z = -3/2, 1/2$ and $S_z = 1/2, -3/2$, respectively. For a system with a more complicated anisotropy, more terms should be added to Equation 2.10.

2.3 Stevens' operators

According to Wigner-Eckart theorem, as a spherical tensor operator, the ZFS Hamiltonian can also be expressed in a more general form by summing the Stevens' operators [7]:

$$\mathscr{H}_{\mathrm{ZFS}} = \sum_{n=0} \sum_{m=0}^{n} B_n^m \widehat{O}_n^m, \tag{2.11}$$

where $\widehat{O}_n^m$ and B_n^m are the Stevens' operators and corresponding coefficients related to anisotropy. The Stevens' operators $\widehat{O}_n^m$ are nth degree polynomials composed of three total angular momentum operators, i.e. either $\boldsymbol{J}_x$, $\boldsymbol{J}_y$, $\boldsymbol{J}_z$, or $\boldsymbol{J}_z$, $\boldsymbol{J}_+$, $\boldsymbol{J}_-$ ($\boldsymbol{J}_\pm = \boldsymbol{J}_x \pm i\boldsymbol{J}_y$). n and m indicate the rank and the component of the spherical tensor, which are related to the quantum number of angular momentum and the order of raising operator $\boldsymbol{J}_+$ and lowering operator $\boldsymbol{J}_-$ respectively. Some of the Stevens' operators vanish in particular conditions [1, 8]. The general rules are listed below.

(i) Regarding time reversal symmetry, the state with opposite angular momentum should be degenerate. Thus only even powers of angular momentum operators remain, odd n terms vanish.

(ii) All terms of $n > 2J$ vanish, where J is the quantum number of total angular momentum under consideration, because the operators only act on $2J + 1$ micro-states.

(iii) Similarly, all terms of $n > 2l$ vanish, where l is the orbital quantum number of the subshell where considered electrons stay. Hence for d and f electrons of an ion, only Stevens' operators up to fourth and sixth degree contribute, respectively.

(iv) Terms also vanish when they are incompatible with the particular point symmetry of the system.

The exact components of each Stevens' operator $\widehat{O}_n^m$ and non-vanishing terms under various point symmetry can be found elsewhere [9, 2, 10]. The corresponding coefficients B_n^m cannot be deduced by group theory or any other analysis. It can only be obtained by *ab initio* calculation or experimental measurement.

The following Hamiltonian composed of Stevens' operators is used in Section 4.1.6 to interpret the spin states of $3d$ ions on single magnetic molecules, which are in the case of $C_{1\mathrm{v}}$ symmetry.

$$\begin{aligned}\mathscr{H}_{\mathrm{eff}} &= B_2^0\widehat{O}_2^0 + B_2^2\widehat{O}_2^2 + B_2^1\widehat{O}_2^1 \\ &= 3B_2^0\boldsymbol{J}_z^2 + \frac{B_2^2}{2}(\boldsymbol{J}_+^2 + \boldsymbol{J}_-^2) + \frac{B_2^1}{4}[\boldsymbol{J}_z(\boldsymbol{J}_+ + \boldsymbol{J}_-) + (\boldsymbol{J}_+ + \boldsymbol{J}_-)\boldsymbol{J}_z].\end{aligned} \tag{2.12}$$

The first two terms are similar to the phenomenological Hamiltonian expressed by Equation 2.10. The last term is a non-vanishing term in $C_{1\mathrm{v}}$ symmetry, which mixes neighbouring original eigenstates and results from a complicated three-dimensional anisotropy. The coefficients B_2^0, B_2^2 and B_2^1 will be determined by fitting the experimental data.

2.4 Spin crossover

As discussed in Section 2.1, the degenerate orbitals in one subshell split into various energy levels in the presence of crystal field. As an typical case, the five d orbitals of an ion in an octahedral complex split into three $\boldsymbol{T_2}$ orbitals and two $\boldsymbol{E}$ orbitals (sketched in Figure 2.1). The energetic order of the $\boldsymbol{T_2}$ and $\boldsymbol{E}$ levels affects the configuration of d electrons for the ground state, i.e. the electrons will not start to occupy the $\boldsymbol{E}$ orbitals unless all of the three $\boldsymbol{T_2}$ orbitals are filled. To respect Pauli exclusion principle, only two electrons are allowed for each orbital at maximum. Therefore, the first six d electrons only occupy the $\boldsymbol{T_2}$ orbitals. However, this concept is incompatible with the Hund's first rule, which requires the state with lowest energy to maximize the total spin quantum number S. Hence, under Hund's first rule, each electron prefers occupying its own orbital with same spin orientation unless there is no empty orbital. Actually, the eventual electron configuration is the result of the competition between crystal field splitting and Hund's first rule. If the system presents a strong crystal field, Hund's first rule will be broken, the electron configuration is dominated by crystal field splitting. Electrons prefer completing the lowest levels before starting to fill the higher levels. Namely, the maximum total spin quantum number S is not guaranteed, as the energy gap Δ of crystal field splitting is too large. This electron configuration is usually labelled as low-spin (LS) state. In comparison, when the strength of crystal field is not sufficient to overcome Hund's first rule, the electron configuration still obeys Hund's first rule. S is maximized by occupying different orbitals with same spin orientation and avoiding sharing one orbital between two electrons with opposite spin. This electron configuration is called high-spin (HS) state. For some systems, these two electron configurations do not show any difference. However, for other systems, e.g. ions with 4-7 d electrons, LS and HS states correspond to dramatically different S and consequent magnetic properties. Apparently, the eventual choice between two electron configurations is largely determined by the strength of the crystal field. Therefore, switching between two electron configurations occurs if the strength of the crystal field can be tuned. This switching effect is named spin crossover (SCO).

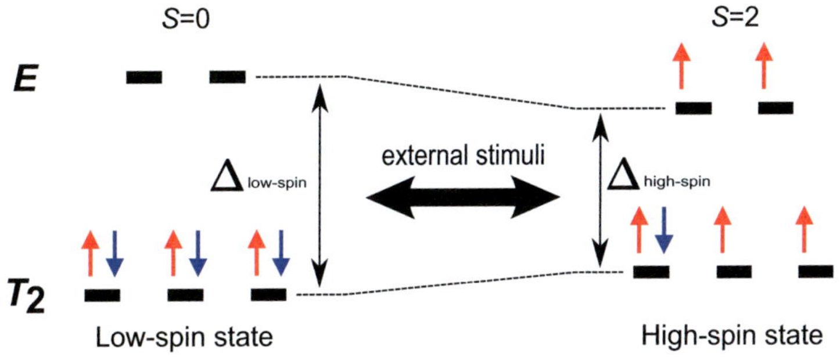

Figure 2.4: Illustration of spin crossover of six $3d$ electrons in octahedral crystal field. Electron configuration presents a LS state ($S = 0$) when crystal field splitting Δ is higher than the electron pairing energy (left case). Otherwise, the HS state ($S = 2$) is preferred (right case). The red upwards (blue downwards) arrows represents spin-up (spin-down) electrons. The length of vertical double-head arrow indicates the strengh of crystal field. The horizontal double-head arrow implies that the switching between LS and HS states can be controlled by external stimuli.

To demonstrate SCO, the case of six $3d$ electrons on an ion in an octahedral complex is sketched in Figure 2.4 as an example. When the complex presents a strong crystal field, i.e. a large energy gap Δ between $\boldsymbol{T_2}$ and $\boldsymbol{E}$ level, the six $3d$ electrons fill the three $\boldsymbol{T_2}$ orbitals. For this low spin state, the total spin quantum number $S = 0$, which means the complex is non-magnetic. Conversely, if the crystal field is relatively weak, Hund's first rule wins the competition and thence dominates the electron configuration. Thus, only one of $\boldsymbol{T_2}$ orbitals will be occupied by a pair of electrons, the rest four orbitals including two from $\boldsymbol{E}$ level will be occupied by remained four electrons individually. In this high spin condition, S is maximized to 2, since the spins are aligned to one direction for last four electrons distributed on four orbitals separately. Consequently, the identical complex but at HS state shows a magnetic moment. More interestingly, the atom can be switched between LS and HS states by controlling crystal field through various external stimuli such as temperature, light, electric and magnetic field or charge flow.

In this dissertation, our investigation of SCO effect on Fe-phen molecules will be presented in Chapter 6.

2.5 Tunnel junction

A tunnel junction refers to an electronic barrier which is composed of two conducting components separated by a thin insulating layer. Classically, if the kinetic energy of an electron in two conductors of the tunnel junction is lower than the barrier height, the electron is absolutely forbidden to pass through the insulating layer. Instead it will be reflected by the barrier. However, from the point of view of quantum mechanics, the wave function of an electron distributes in the whole junction. Therefore, there is always a probability for electron to go through the barrier and reach the other side. This effect is called quantum tunneling. It is widely observed in natural phenomena, such as nuclear fusion in stars and radioactive decay. Tunnel junction has also been extensively used in applications, for instance, magnetic tunnel junction in the head of modern hard disk drives; superconducting tunnel junction (Josephson junction) in superconducting quantum interference devices (SQUID) and consecutive $p-n$ junctions in photovoltaic cells. On the other hand, quantum tunneling effect also limits the density of integrated circuits by causing current leakage. Scanning tunneling microscopy, the main experimental technique used in this dissertation, also has at its heart a tunneling junction.

2.5.1 One-dimensional rectangular barrier

To understand quantum tunneling, the concepts of Heisenberg uncertainty principle and wave-particle duality must be employed. In the following, the model of a rectangular potential barrier in one dimension will be discussed as an simplest example of tunnel junction.

As we know, from the point of view of wave-particle duality, an electron can also be described by a wavefunction $\boldsymbol{\Psi}(\boldsymbol{r})$. The squared modulus $|\boldsymbol{\Psi}(\boldsymbol{r})|^2$ describes the probability density of the electron being located at the position $\boldsymbol{r}$. In one dimension, the corresponding wave function of an electron $\boldsymbol{\Psi}(x)$ follows the time-independent Schrödinger equation:

$$-\frac{\hbar^2}{2m}\frac{d^2}{dx^2}\boldsymbol{\Psi}(x)+\boldsymbol{V}(x)\boldsymbol{\Psi}(x)=E\boldsymbol{\Psi}(x), \tag{2.13}$$

where $\hbar$, m and E are the reduced Planck's constant, the mass and the total energy of the electron. $\boldsymbol{V}(x)$ is the potential energy at position x, which is equal to U_0 between $x=0$ and $x=a$ and zero elsewhere (see Figure 2.5). Thus, the x axis is divided into three sections: one barrier region from $x=0$ to $x=a$ with a finite potential U_0; two sides of the barrier with potential $U=0$ refer to two conductors. Since the potential is constant for every section, the electron can be regarded as a free particle in each section, thus the solution of Equation 2.13 can be expressed as a superposition of left and right moving plane waves, respectively:

$$\begin{aligned}
\Psi_{\mathrm{L}}(x) &= A_{\mathrm{r}}e^{ik_0x}+A_{\mathrm{l}}e^{-ik_0x} && x<0;\\
\Psi_{\mathrm{C}}(x) &= B_{\mathrm{r}}e^{ik_1x}+B_{\mathrm{l}}e^{-ik_1x} && 0<x<a;\\
\Psi_{\mathrm{R}}(x) &= C_{\mathrm{r}}e^{ik_0x}+C_{\mathrm{l}}e^{-ik_0x} && x>a.
\end{aligned} \tag{2.14}$$

Here, k_0 and k_1 are the wave numbers for the sections with different potential:

$$\begin{aligned}
k_0 &= \sqrt{2mE}/\hbar && x<0 \quad \text{or} \quad x>a,\\
k_1 &= \sqrt{2m(E-U_0)}/\hbar && 0<x<a.
\end{aligned} \tag{2.15}$$

The coefficients A, B and C can be figured out by considering the boundary conditions, which hint that both the wave function and its derivative must be continuous on the edges of the barrier. Assuming the plane wave electron comes from the left side of the barrier, the left moving wave $C_{\mathrm{l}}e^{-ik_0x}$ on the right of the barrier can be omitted since there is no wave coming from right. Note, when the barrier is higher than the energy of the electron (i.e. $U_0>E$), k_1 becomes imaginary, i.e., Ψ_{C} is an exponentially decaying function rather than a wave. Under these restrictions on the coefficients, the transmission probability T for electron to tunnel through the barrier can be described as:

$$T=|\frac{C_{\mathrm{r}}}{A_{\mathrm{r}}}|^2=\frac{1}{1+\frac{U_0^2}{4E(U_0-E)}\sinh^2(\frac{\sqrt{2m(U_0-E)}}{\hbar}a)}. \tag{2.16}$$

Note that this probability never goes to zero, which means the electron can always tunnel through the barrier. However, this probability only becomes significant under particular conditions. It requires that the energy of electron E is comparable with the height of the barrier U_0, moreover, the thickness of the barrier a should not be much larger than the wave length of the electron $2\pi\hbar/\sqrt{2mE}$. For a thin barrier ($\frac{\sqrt{2m(U_0-E)}}{\hbar}a \gg 1$),

$$T \approx \frac{16E(U_0-E)}{U_0^2}\exp\left(\frac{-2\sqrt{2m(U_0-E)}}{\hbar}a\right), \tag{2.17}$$

which implies that the transmission probability exponentially decays with increasing barrier thickness.

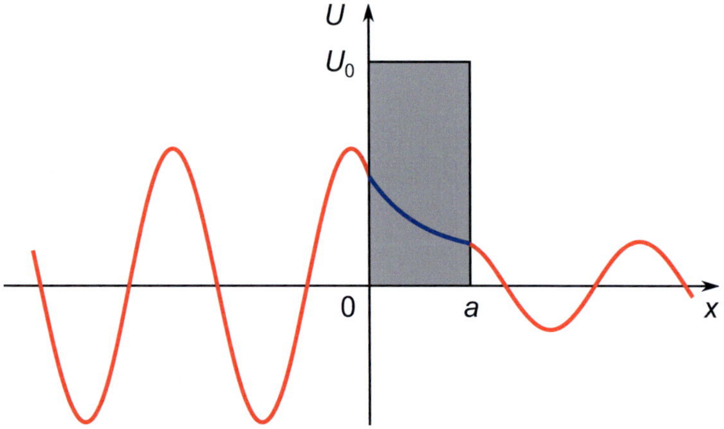

Figure 2.5: Sketch of quantum tunneling in the case of one-dimensional rectangular barrier. U_0 and a indicate the height and thickness of the barrier (grey area). The incident plane wave from left side partially crosses the barrier and arrives the right side with a reduced amplitude.

Although the one-dimensional rectangular barrier is a simple quasi-classical model, it illustrates the principle of quantum tunneling vividly. It is regarded as a suitable model for various tunnel junctions, including STM, where electrons cross the vacuum gap between a metallic tip and a conducting surface to produce a tunneling current. Accoriding to Equation 2.17, the tunneling current depends exponentially on the distance between the tip and surface. So, by adjusting the tunneling current, the distance between the tip and surface can be precisely controlled, and vice versa. This way the topography of the surface can be obtained by STM. More technical details will be introduced in Section 3.3.

However, the actual potential is not homogeneous over the surface. Therefore, the tunneling current still varies even when the tip-sample distance is kept constant when the tip moves over the surface. That is why the locations of atoms can be reconstructed by monitoring the tunneling current in an atomic terrace. This diversity of potential on the surface corresponds to the local density of states (LDOS, for detail see next section). To understand its contribution to tunneling current, the model of one-dimensional rectangular barrier is not enough. One possible solution is Bardeen's approach (see next section).

2.5.2 Bardeen's approach

The real cases (e.g. STM) are much more complicated than the model of one-dimensional rectangular barrier. First, the quantum tunneling is in three dimensions rather than one. Moreover, the barrier is not rectangular any more. It strongly depends on the work functions of both electrodes (referring to the tip and the sample in the case of STM). Additionally, the electrons in both electrodes cannot move as free particles. They follow the electronic band structure. To explain the tunneling current in an actual junction, J. Bardeen proposed an approach from a view of many-particle physics [11]. After the invention of STM, this method was extended to fit the case of STM [12, 13].

Under Bardeen's theory, the junction is treated as two parts rather than as a single system, although there is an overlap of the electronic states between two separated electrodes. As expressed below, for an electron, the transmission probability from a tip state Ψ_t to a sample state Ψ_s under first-order perturbation theory is governed by Fermi's golden rule.

$$T_{ts} = \frac{2\pi}{\hbar}|M_{ts}|^2\delta(E_t - E_s), \tag{2.18}$$

where $\delta(E_t - E_s)$ is a delta function which only permits elastic tunneling ($E_t = E_s$), all of the inelastic tunneling is not taken into account in Bardeen's approach. The transition matrix M_{ts} strongly depends on the geometric and electronic properties of both sample and tip. It can be calculated by integrating the overlap between the wave functions in tip and sample. The detail of the related derivation is given in Ref [13].

The current induced by electron transmission through the channel from Ψ_t to Ψ_s can be calculated by weighting the transmission probability with the occupation probability of Ψ_t and the probability that Ψ_s is unoccupied, which are described by the Fermi-Dirac distribution at finite temperature,

$$i_{ts} = -\frac{4\pi e}{\hbar} f(E_t)[1 - f(E_s)]|M_{ts}|^2\delta(E_t - E_s), \tag{2.19}$$

where the Fermi-Dirac distribution writes $f(E) = [\exp(\frac{E-E_\mathrm{F}}{k_\mathrm{B}T}) + 1]^{-1}$. Note that the factor $-2e$ that accounts for the charge and spin multiplicity of each electron. Three possible tunneling channels are sketched in Figure 2.6. The total tunneling current from the sample to the tip is the difference of the sums of electron transmission between all states in two opposite directions, which can be calculated as:

$$\begin{aligned} I = & \frac{4\pi e}{\hbar}\sum_{\mathrm{t,s}} f(E_\mathrm{t})[1-f(E_\mathrm{s})]|M_\mathrm{ts}|^2\delta(E_\mathrm{t}-E_\mathrm{s}) \\ & -\frac{4\pi e}{\hbar}\sum_{\mathrm{s,t}} f(E_\mathrm{s})[1-f(E_\mathrm{t})]|M_\mathrm{st}|^2\delta(E_\mathrm{s}-E_\mathrm{t}) \\ = & \frac{4\pi e}{\hbar}\sum_{\mathrm{t,s}} [f(E_\mathrm{t})-f(E_\mathrm{s})]|M_\mathrm{ts}|^2\delta(E_\mathrm{t}-E_\mathrm{s}). \end{aligned} \tag{2.20}$$

Here, the transition matrix is supposed to be symmetric, i.e. $|M_\mathrm{ts}| = |M_\mathrm{st}|$. Assuming a continuous energy spectrum, one can express the tunneling current as an integral over the density of states (DOS) of the tip (ρ_t) and the sample (ρ_s):

$$\begin{aligned} I = & \frac{4\pi e}{\hbar}\int_{-\infty}^{\infty}[f(E_\mathrm{F}+\varepsilon)-f(E_\mathrm{F}+V+\varepsilon)]\times \\ & \rho_\mathrm{t}(E_\mathrm{F}+\varepsilon)\rho_\mathrm{s}(E_\mathrm{F}+eV+\varepsilon)|M(V,\varepsilon)|^2\,d\varepsilon, \end{aligned} \tag{2.21}$$

where V is the bias voltage applied on the sample with respect to the tip. Usually $\rho_\mathrm{t}(E)$ can be treated as a constant, while $\rho_\mathrm{s}(E)$ varies with the position on the sample, which is the most important information hiding behind the tunneling current.

From Equations 2.17 and 2.21, we learned that not only the tip-sample distance changes the tunneling current exponentially, but also the DOS of the sample influences the tunneling current dramatically. As STM is a local measurement method, the LDOS of the sample can be mapped by comparing the tunneling current at different points during scanning the tip over the surface with a constant tip-sample distance. The detail about how to extract LDOS from tunneling current will be introduced in next section.

2.6 Scanning tunneling spectroscopy

As discussed in previous section, the tunneling current contains information on the DOS. Actually, by measuring and analysing the tunneling current, a series of interesting physical properties of the sample can be revealed. Besides to obtain the position dependent tunneling current under a particular bias voltage, another dominant function for STM is to record the tunneling current at various bias voltages with a fixed tip-sample distance at a particular position on the sample. Since the tunneling current is measured as a function of bias voltage that is directly related to the energy of electron, this operation mode is named scanning tunneling spectroscopy (STS). It is usually employed to investigate the local electronic structure of various materials on the point where the tip is located (i.e. LDOS).

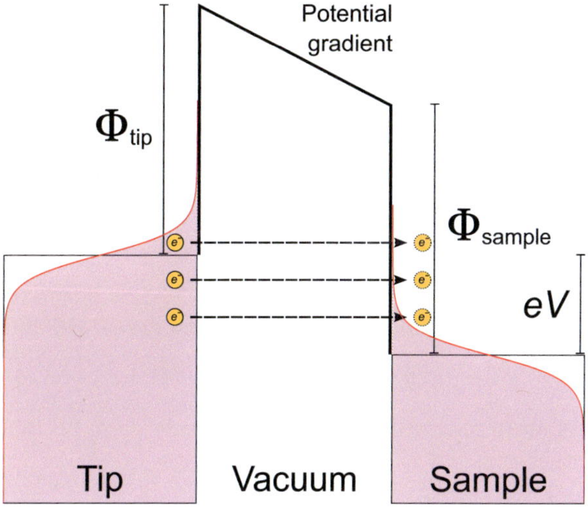

Figure 2.6: Schematic principle of quantum tunneling under Bardeen's approach in the case of STM. The rectangles indicate the states below Fermi levels. The red curves represent the Fermi-Dirac distributions. The pink areas indicate the occupied states at an finite temperature. Dashed arrows indicate electrons crossing the barrier via three channels from an occupied state on one side to an unoccupied state on other side without losing energy. Φ indicates the work function of the tip or sample. A positive bias V is applied on the sample with respect to the tip.

A simple and natural spectrum that can be measured with STS is an $I(V)$ curve, which presents the dependence of the tunneling current on the bias voltage. A more advanced spectrum measures the dependence of the first and even second derivative of tunneling current (dI/dV and d^2I/dV^2) on bias voltage. In contrast to a direct $I(V)$ curve, dI/dV and d^2I/dV^2 curves frequently show physical features more distinctly, especially, dI/dV curves directly reflect the LDOS of the sample as shown below.

From Equation 2.21, we learn that the LDOS of the sample is closely related to the tunneling current. When k_BT is much smaller than the energy resolution of the measurement, the corresponding Fermi-Dirac distribution approaches to a step function. The tunneling current can be further approximated as follows:

$$I = \frac{4\pi e}{\hbar}\int_{-eV}^{0} \rho_{\mathrm{t}}(E_{\mathrm{F}}+\varepsilon)\rho_{\mathrm{s}}(E_{\mathrm{F}}+eV+\varepsilon)|M|^2\,d\varepsilon. \tag{2.22}$$

Here, $|M|$ is the transition matrix element treated as a constant. Further, for small voltage ranges we can assume a constant DOS for the tip. Thus, the derivative of tunneling current with respect to bias voltage is proportional to the LDOS of the sample:

$$\frac{dI}{dV} \propto \rho_{\mathrm{s}}(E_{\mathrm{F}}+eV). \tag{2.23}$$

Until now, we learned that measuring the first derivative of tunneling current is a very convenient method to obtain LDOS of the sample. Via adjusting the bias voltage V applied on the sample, the LDOS of the sample in a reasonable energy range can be plotted. If a positive V is applied, the unoccupied states above Fermi level can be investigated. A negative V corresponds to the information about occupied states below Fermi level. For instance, if a molecule adsorbed on a surface is regarded as the other electrode of the tunnel junction, the energy of the highest-occupied molecular orbital and the lowest-unoccupied molecular orbital can be determined by measuring a dI/dV spectrum.

Further technical details of measuring STS will be introduced in Section 3.3.2.

2.6.1 Local density of states

The DOS and the LDOS have been mentioned many times in the last sections. It is necessary to introduce the physical meaning behind these two concepts. In condensed matter physics, the density of states refers the amount of states per interval of the energy on as a function of the energy. In other words, it describes the distribution of states in the scale of energy. A larger value of the DOS means there are more states at corresponding energy for particles to occupy. If there are no states at a particular energy, the DOS is equal to zero. The DOS is a widely used concept to interpret some physical phenomena of quantum mechanics in theory and calculation. The LDOS is used to describe the local variation of DOS. Because of its extremely high spatial resolution, STM is a perfect instrument to study the LDOS of electron even at the sub-atom scale.

2.6.2 Elastic and inelastic electron tunneling spectroscopy

All of the previous discussion about quantum tunneling only considered elastic processes, which mean the electrons do not lose any energy during tunneling through the barrier. As sketched by the dashed arrows in Figure 2.7, in the case of a positive bias voltage V, the tunneling electrons reach the other side of the barrier with energies higher than the Fermi level thence occupy the corresponding empty states before loosing the additional energy by relaxing to the Fermi level of the sample. If the LDOS near the Fermi level of the sample is assumed to be constant in the energy range involved, from Equation 2.22 it is deduced that the tunneling current is directly proportional to the bias voltage (see Figure 2.8(a)) since a larger V corresponds to more empty states of the sample for the tunneling electrons. Of course, a linear $I(V)$ curve results in a trivial constant first derivative that is consistent with the assumption of the constant LDOS. The same is true for the case of negative V, where electrons tunnel from the occupied states on the sample side.

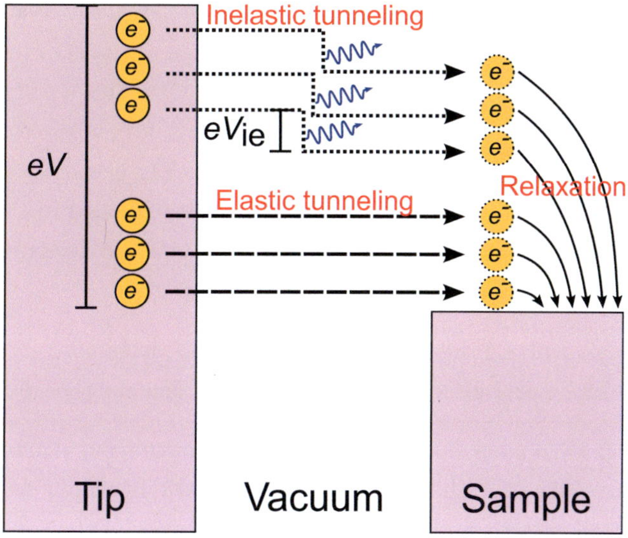

Figure 2.7: Sketch of elastic (dashed arrows) and inelastic (dotted arrows) processes in quantum tunneling. Blue waves refer to quasi-particles with energy eV_{ie} excited in inelastic processes. A positive bias V is applied on the sample.

In comparison to elastic tunneling processes, inelastic ones (as sketched by dotted lines in Figure 2.7) refer to tunneling processes, where electrons interact with other quasi-particles including plasmons, phonons, spin-excitations and so on in the barrier. Frequently, the interaction will cause a transfer of energy. The tunneling electron sacrifices its energy to excite quasi-particles or gains energy from the adsorbate, if the temperature allows. The latter one will not be discussed in this dissertation.

Inelastic tunneling opens a second channel for electrons to cross the barrier via exciting quasi-particles, which results in an additional contribution to the tunneling current. In other words, inelastic excitations offer another transmission possibility for tunneling electrons. But this only starts to take place when the bias voltage reaches the essential threshold V_{ie}, which is determined by the excitation energy of the quasi-particles, because the incident electron needs an additional energy of at less eV_{ie} to excite the quasi-particle (see blue waves in Figure 2.7). Assuming the transmission probability related to inelastic excitation is independent on bias voltage, the part of the current contributed by the inelastic tunneling should be proportional to the amount of electrons with kinetic energy larger than eV_{ie} in the barrier. Therefore, as shown on the $I(V)$ curve in Figure 2.8(b), the current is still linear to the voltage but the slope increases when voltage reaches V_{ie}. Thus, a pair of kinks appears exactly at $\pm V_{\mathrm{ie}}$. Regarding the dI/dV spectrum, symmetric step functions at $\pm V_{\mathrm{ie}}$ result from the combination of original constant LDOS of sample and abrupt appearance of a second tunneling channel (see Figure 2.8(c)). Because of the step function, a pair of delta function arise at $\pm V_{\mathrm{ie}}$ in d^2I/dV^2 spectrum as presented in Figure 2.8(d), where the contribution of elastic tunneling is zero. As the signal of inelastic tunneling is usually smaller than that of elastic tunneling, the positions of the peaks of the delta function in d^2I/dV^2 spectrum indicate the threshold V_{ie} better than the step function in dI/dV spectrum. So, d^2I/dV^2 spectra are frequently used to extract the excitation energy eV_{ie}.

The spectroscopy involving inelastic excitation is called inelastic electron tunneling spectroscopy (IETS) in contrast to trivial elastic electron tunneling spectroscopy.

Actually, the real IETS does not present ideal step functions or delta functions, since the spectroscopic peaks of d^2I/dV^2 not only have intrinsic linewidth but also are always broadened in actual experiments [14]. The intrinsic linewidth at half-maximum $\Gamma_{\mathrm{intrinsic}}$ can be ascribed to the natural energy broadening which is linked to the finite lifetime by the Heisenberg uncertainty principle. Moreover, thermal broadening and instrumental broadening usually contribute to the visible width of the peak. The former one results from the smearing of Fermi level caused by Fermi-Dirac distribution at non-zero temperature. The corresponding half-width predicted by Lambe and Jaklevic [14] is $\Gamma_{\mathrm{thermal}} = 5.4k_{\mathrm{B}}T$. It is impossible to avoid the second broadening, when a finite modulation is applied to the bias voltage when using a lock-in technique to measure the derivatives (see Section 3.3.2). The modulation induced broadening can be phenomenally explained by the fluctuation of the Fermi level of the sample under an oscillating bias voltage. Klein and co-authors gave the linewidth of modulation broadening as $\Gamma_{\text{modulation}} = 1.22\Delta V$ [15], where ΔV is the amplitude if the modulation voltage. Basing on the above there kinds of typical energy broadening, the eventual half-width at half-maximum of inelastic excitations in the experimental d^2I/dV^2 spectrum can be described as follows:

$$\begin{aligned}\Gamma_{\text{experiment}} &= \sqrt{\Gamma^2_{\text{intrinsic}} + \Gamma^2_{\text{thermal}} + \Gamma^2_{\text{modulation}}} \\ &= \sqrt{\Gamma^2_{\text{intrinsic}} + (5.4k_BT/e)^2 + (1.22\Delta V)^2}.\end{aligned} \tag{2.24}$$

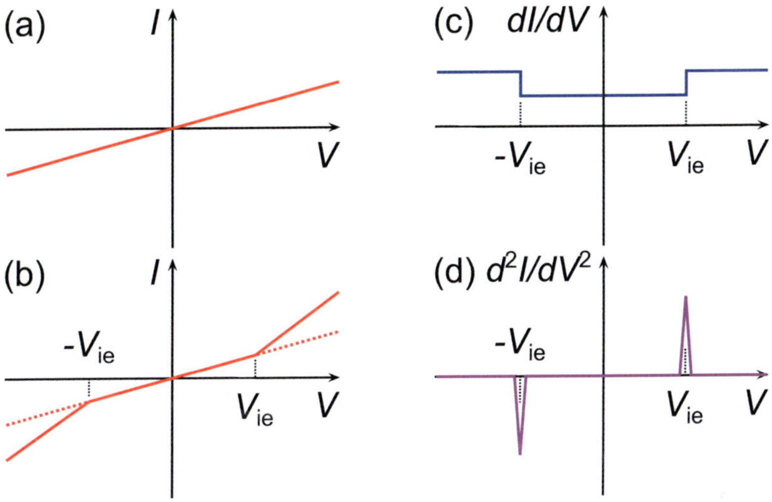

Figure 2.8: Schematic plots for elastic and inelastic electron tunneling spectroscopy. (a) and (b) refer to the $I(V)$ curves for pure elastic tunneling and inelastic tunneling joined, respectively. (c) and (d) are the corresponding dI/dV and d^2I/dV^2 spectra. Note that the inelastic tunneling starts to take place at a bias voltage exceeding V_{ie} at both polarities.

Note that the thermal broadening is independent on instrument or measuring technique, it sets a limit for the energy resolution. The only way to lower this limit is decreasing the temperature. To guarantee that the thermal broadening does not mask the investigated excitations, most STS is taken at cryogenic temperatures. Regarding IETS, the inelastic excitation is detectable only when Γ_{thermal} is not much larger than the excitation energy. Additionally, according to Equation 2.24, the energy resolution can be improved by decreasing the modulation voltage. However, that will reduce the ratio between signal and noise. The details about adjusting the parameters of lock-in will be discussed in Section 3.3.2.

2.7 The Kondo effect

The electrical resistivity decreases monotonously with lowering temperature in a metal because of less scattering due to suppressed lattice vibrations. The resistivity of metal containing magnetic impurities, however, increases again when temperature decreases below a particular point. This physical phenomenon is called the Kondo effect to honour Jun Kondo, who first described this effect theoretically in 1964 as the scattering of the conduction electrons due to the magnetic impurities with the help of third order perturbation theory [4], after the minimum resistivity has been observed for many times since 1930s [16, 17].

2.7.1 Minimum resistivity under Kondo model

Kondo described the Hamiltonian of a single impurity interacting with conduction electrons as follows [4]:

$$\mathscr{H} = \sum_{k,s} \varepsilon_k a^\dagger_{ks} a_{ks} - J \sum_{k,k'} (a^\dagger_{k'\uparrow} a_{k\uparrow} - a^\dagger_{k'\downarrow} a_{k\downarrow}) S_z - J \sum_{k,k'} (a^\dagger_{k'\uparrow} a_{k\downarrow} S^- + a^\dagger_{k'\downarrow} a_{k\uparrow} S^+), \tag{2.25}$$

where the first term refers to the unperturbed Hamiltonian expressing the energy of all conduction electrons, the second and third terms are the perturbing Hamiltonians describing the scattering when the spin of the impurity is conserved and the scattering when the spin flips, respectively. J indicates the strength of spin exchange interaction between the impurity and conduction electrons.

Kondo summarized a phenomenological expression for the temperature dependent resistivity of metal with magnetic impurities [4]:

$$\rho(T) = \rho_0 + aT^5 + bJln(\frac{T}{c}), \tag{2.26}$$

where a, b and c are constants, ρ_0 is the residual resistivity in the metal, aT^5 results from lattice vibrations. The last logarithmically dependent term is ascribed to the scattering happening around the impurities, which was first derived by Kondo from Equantion 2.25. Note, the value of last term increases as the temperature decreases, when J is negative which implies that the spin interaction between the conduction electrons and local impurities is anti-ferromagnetic. So, this model offers a solution to the problem of the minimum resistivity of metal containing magnetic impurities at a low temperature.

However, the logarithmic term diverges at zero temperature. Actually, Kondo's perturbation theory fails to deal with the strong coupling at low temperatures around the so-called Kondo temperature T_K, which is related to the location of the minimum resistivity. By using non-perturbative rather than perturbative techniques, the combination of Anderson impurity model and Wilsonian re-normalization theory was developed to understand the physics of the Kondo effect better [18, 19]. This advanced model has the ability to solve the Kondo problem at temperatures close to and below T_K, which corresponds to a regime of strong coupling. It offers a finite resistivity at zero temperature but still retains the minimum resistivity. Under this model, the local magnetic impurity is screened by conduction electrons through an anti-ferromagnetic interaction. The screening effect increases the probability of conduction electrons to be scattered, which raises the resistivity. As temperature approaches zero, the spins of the impurity and the screening electrons form a many-body non-magnetic singlet state around the Fermi level.

2.7.2 Kondo resonances in the Anderson model

Besides the temperature dependent resistivity, the Kondo effect is manifested in spectroscopic measurements, which presents a low-energy excitation from the ground state to the many-body non-magnetic singlet state. This excitation with a characteristic width of $\Gamma = k_B T_K$ is named Kondo resonance. It can be detected by high-resolution photo-electron spectroscopy [20, 21], however, the signal is averaged over a surface area of the size of the beam in the range of 1 mm^2 typically. Alternatively, tunneling spectroscopy offers another way to directly detect the Kondo resonance in the case of a single magnetic impurity embedded in a tunnel junction. During the last two decades, this new method has opened another door to Kondo physics. Experimentally, measuring Kondo resonance with tunneling spectroscopy can be realized on the magnetic quantum dots in single electron transistors [22], or above single magnetic atom or ion on metallic surface in scanning tunneling junction [23]. Obviously, the latter one is an important research field to study Kondo resonance locally, because STM not only has a high spatial resolution but can also manipulate single atoms precisely. More information of the Kondo resonance will be introduced in the following.

As mentioned before, the Kondo resonance corresponds to the excitation to the many-body non-magnetic singlet state. In a little bit more detail, it can be briefly explained as following [24]. As described under the Kondo model, the local spin of the impurity flips between $\uparrow$ and $\downarrow$ states via scattering with the conduction electrons, while under the Anderson model, the flipping process simultaneously creates a spin-excitation state around Fermi level. To understand that, the electronic structure of the magnetic impurity embedded in a host metal is simplified to a single-orbital state (usually $3d$ or $4f$ orbital), which means the state can be occupied by zero, one or two electrons. Taking the $4f$ orbital as an example, in order to be magnetic, the ground state must be singly occupied state $4f^1$, which is located below Fermi level with a binding energy $-\varepsilon$, while the $4f^2$ state filled by two electrons with opposite spins costs an additional energy U due to Coulomb repulsion as sketched in Figure 2.9(a). As a result of the hybridization between the $4f$ states and electronic states from the host metal, the energy levels of $4f^1$ and $4f^2$ are broadened by $\Delta = \rho_0 |V|^2$, where ρ_0 is the DOS of the metal at Fermi level, V is the hybridization matrix element. As schematically presented in Figure 2.9(b), the spin-flip process of the electron on the ground state $4f^1$ can be achieved via two ways: either first exciting the single $4f$ electron to the Fermi level of metal ($4f^1 \rightarrow 4f^0$) subsequently filling the $4f$ orbital with a conduction electron carrying an opposite spin ($4f^0 \rightarrow 4f^1$), or first filling the orbital with another conduction electron ($4f^1 \rightarrow 4f^2$) followed by exciting the initial occupying electron ($4f^2 \rightarrow 4f^1$). In a classical interpretation, these two ways require ε and $\varepsilon + U$ to overcome the barriers respectively, although the total energy is conserved. However, under the principles of quantum mechanics, the Heisenberg uncertainty guarantees this spin exchange as a virtual process with a short excitation time, during which a conduction electron is involved. Due to the large amount of these virtual processes caused by conduction electrons, the localized $4f$ electron of impurity partially occupies the states on Fermi level of the metal, in turn, the conduction electrons get a chance to stay on the $4f$ orbital. Thus, a new state is generated close to the Fermi level, which is a typical many-body phenomenon.

Obviously, as an electronic state, the Kondo resonance can be detected by STS. As sketched in Figure 2.9(c), there are three paths for electrons to tunnel through the vacuum from the tip to the sample. The direct transmission into the unoccupied state of the sample (path 1) and the transmission into the Kondo resonance (path 2) interfere with each other at Fermi level of the metal. Consequently, a characteristic feature is produced in the differential conductance (dI/dV) spectrum. While the process of path 3 does not contribute to the interference because of the opposite spin of the fine state.

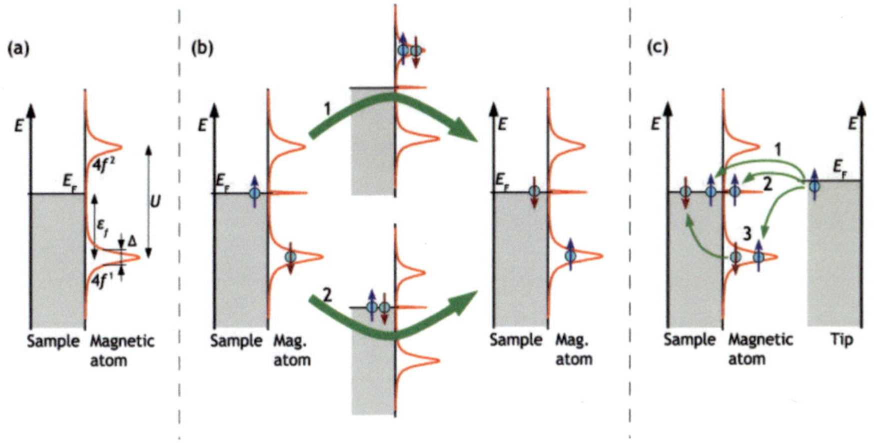

Figure 2.9: Sketch of the formation of the Kondo resonance under Anderson model. (a) The energy distribution of states of a single magnetic atom with a single $4f$ orbital located close to the Fermi level of the host metal. (b) The Kondo resonance near Fermi level results from a spin-flip process for the electron in the singly occupied 4f state, caused by a conduction electron with opposite spin from the metal, which can be achieved via two virtual ways indicated by two green arrows: either doubly occupying (path 1) or emptying (path 2) the $4f$ orbital as the intermediate step. (c) The tunneling electrons from the probing tip have three channels to reach the sample: elastic tunneling to the unoccupied state on the Fermi level of sample (1); to the Kondo resonance state of the magnetic atom (2) or to the $4f$ orbital of the magnetic atom indirectly via a spin-flip process (3). This figure is reprinted with permission from Ref. [24].

At zero temperature, the half-width at half-maximum Γ_0 of the Kondo resonance, which directly indicates the Kondo temperature T_K, can be expressed by the parameters of the system introduced before [25]:

$$\Gamma_0 = k_\mathrm{B} T_\mathrm{K} \simeq \sqrt{2\Delta \frac{U}{\pi}} \exp[-\frac{\pi}{2\Delta(|\frac{1}{\varepsilon}| + |\frac{1}{\varepsilon+U}|)}]. \tag{2.27}$$

This equation reveals that the width of Kondo resonance and the Kondo temperature decisively depend on the energy structure of the single orbital of the impurity. Similar to the thermal broadening discussed in Section 2.6.2, at a non-zero temperature T, the width of Kondo resonance is broadened to $\Gamma = \sqrt{(\alpha k_\mathrm{B} T)^2 + (2k_\mathrm{B} T_\mathrm{K})^2}$ [26, 27]. In the $T \gg T_\mathrm{K}$ regime, the width depends linearly on the temperature, the parameter α approaches 5.4, which is determined by the Fermi-Dirac distribution. At the same time, the height of the Kondo resonance also presents a strong temperature dependence, which can be described as $1 - c(T/T_\mathrm{K})^2$ in $T \ll T_\mathrm{K}$ regime, it logarithmically decays as temperature increases for $T > T_\mathrm{K}$.

In $T \ll T_K$ regime, since the magnetic impurity stays at a screening induced non-magnetic singlet state, a small external magnetic field cannot disturb the Kondo effect. As magnetic field increases, the Kondo resonance starts to split and eventually is suppressed completely and replaced by a pair of inelastic excitation peaks as the result of Zeeman splitting [28]. The influence of magnetic field becomes noticeable when the Zeeman energy is comparable to the Kondo temperature, i.e. $B \sim \frac{k_B T_K}{2g\mu_B}$ [29].

The temperature and magnetic field dependences are typically used as evidence to confirm the existence of Kondo resonance in STS experiment.

However, in actual dI/dV spectra, the Kondo resonance cannot be measured independently, since it is always embedded in the elastic tunneling background. The interference between them results in a particular feature depending on the electronic characteristics of the metal and the magnetic impurity. Furthermore, even in a same system, the temperature, vertical height and lateral position of the tip alter the feature a lot. To fit the feature of Kondo resonance in STS, a Fano function is the widely used, which will be introduced below.

2.7.3 Fitting Kondo resonance with a Fano function

In physics, the interference between a resonant scattering process and a background of elastic process gives rise to an asymmetric lineshape in spectroscopy. This phenomenon is called Fano resonance to honour Ugo Fano, who contributed the first theoretical interpretation in the case of the inelastic scattering of electrons from helium [30]. As a general phenomenon of waves, Fano resonances occur widely in various fields. Obviously, the Kondo resonance in STS can also be regarded as a Fano resonance.

In the Fano model, a tunneling electron can tunnel either to the discrete state on the magnetic impurity (resonant process) or to continuum states in the metal (elastic process). The total transmission rate for an electron with energy ε from an arbitrary initial state in the tip to any final state in the sample can be expressed as:

$$R(\varepsilon) = R_0(\varepsilon)\frac{(q+\varepsilon')^2}{1+\varepsilon'^2}, \qquad \varepsilon' = \frac{\varepsilon - \varepsilon_0}{\Gamma}, \tag{2.28}$$

where ε_0 is the energy of the discrete state, Γ is the half-width at half-maximum of the Kondo resonance, R_0 refers to the transmission rate without discrete state, and the so-called Fano parameter q indicates the ratio between the matrix elements corresponding to the transmissions to the discrete and continuum states, respectively. The absolute value and sign of q depend on the relative strength and phase of the resonant process compared to the elastic process. Note that it is the variation of the phase that generates the asymmetry of the line-shape. This equation describes the line-shape of a Fano function. It is frequently used to fit the Kondo resonance in STS, consequently parameters of the Kondo effect can be extracted.

It was found that the energy parameter ε' can be written as following in the regime of temperature lower than T_K [31]:

$$\varepsilon' = \frac{\varepsilon - \alpha}{k_B T_K}, \tag{2.29}$$

where α is a constant. The above equation reveals that the linewidth of Kondo resonance is directly linked to the Kondo temperature by Equation 2.27.

The Fano function is applied to fit the Kondo resonances we observed on single magnetic molecules in Sections 4.1.3 and 4.2.1.

2.8 Magneto-optic Kerr effect

In 1877, John Kerr discovered the magneto-optic Kerr effect (MOKE), which refers to the alternation of light after the reflection from a magnetic surface [32]. The alternation can be the change of either the polarization or the intensity. This magneto-optic effect is very similar to the Faraday effect, which describes the changes of transmitted light instead of reflected one. The physics of MOKE can be interpreted by the combination of quantum mechanics and classical optics [33]. Simply speaking, the light interacts with the magnetic sample, but the left- and right-handed circularly polarized waves experience different interactions. Thus, the polarization of the light changes after reflection.

2.8.1 Geometries

In the experiment, due to the different orientations of the magnetization of the sample with respect to the incidence plane and reflecting surface, MOKE set-ups use different geometries to yield optimal contrast of the Kerr signal from the detecting light in order to study the corresponding magnetizations with a high precision. In the following, three frequently used geometries, polar, longitudinal and transversal MOKE will be introduced.

Polar MOKE

The polar MOKE is frequently used to study the out-of-plane magnetization of a sample (see Figure 2.10(a)). To maximize the Kerr effect, linearly polarized incident light is set perpendicular to the magnetic surface in the case of polar MOKE. The complex reflectivity r of the right- $(+)$ or left-handed $(-)$ circularly polarized light can be expressed as

$$r_{\pm} = \frac{E^r_{\pm}}{E^i} = |r_{\pm}|e^{i\Delta_{\pm}}, \tag{2.30}$$

where E^i and $E^r_{\pm}$ represent the complex amplitude of the incident light and reflected light in right- $(+)$ or left-handed $(-)$ circularly polarized case, respectively. In comparison to the incident light, the reflected light not only presents a magnetization dependent rotation of the polarization plane, it but also turns to be elliptically polarized.

The Kerr rotation θ_{K} and ellipticity ε_{K} can be written as

$$\theta_{\mathrm{K}} = -\frac{1}{2}(\Delta_{+} - \Delta_{-}), \tag{2.31}$$

$$\varepsilon_{\mathrm{K}} = \frac{|r_{+}| - |r_{-}|}{|r_{+}| + |r_{-}|}. \tag{2.32}$$

The complex Kerr angle is expressed as

$$\Phi_{\mathrm{K}} = \theta_{\mathrm{K}} + i\varepsilon_{\mathrm{K}}. \tag{2.33}$$

Note both θ_{K} and ε_{K} are proportional to the out-of-plane component of the magnetization.

Longitudinal MOKE

In the longitudinal geometry, the sample presents an in-plane magnetization, which is simultaneously parallel to the plane of the incidence as sketched in Figure 2.10(b). Unlike in the case of polar MOKE, the incident light is not normal to the surface in the longitudinal set-up. A p-polarized light (linearly polarized light with $\boldsymbol{E}$ vector parallel to the plane of incidence) is employed as the incident light. After reflection, the detecting light becomes elliptically polarized, in other words, a s-polarized component arises in the reflected light. Experimentally, the change of the intensity of the reflected light is usually measured [34], which is directly proportional to the component of the magnetization that is parallel to both the sample surface and the plane of incidence.

Transversal MOKE

The transversal geometry corresponds to the case that the in-plane magnetization is perpendicular to the plane of incidence (see Figure 2.10(c)). In this case, the detecting light is also not perpendicular to the magnetic surface. However, unlike in two previous geometries, the signal of transversal MOKE results from the change of the reflectivity r instead of the polarization. As above, the change of r is proportional to the component of the magnetization that is parallel to the sample surface but perpendicular to the plane of incidence.

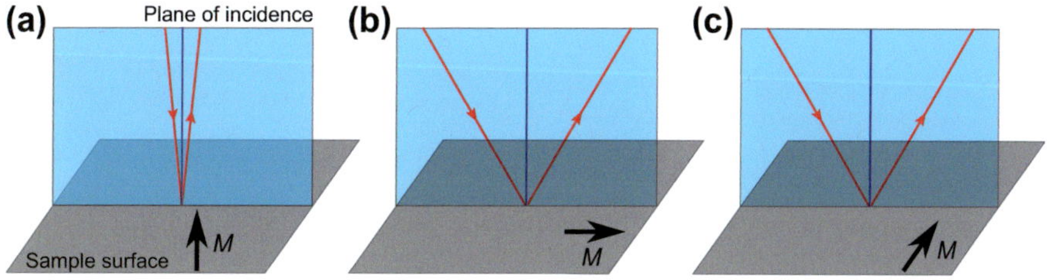

Figure 2.10: Sketch of three tyies of geometries for MOKE measurement. (a) polar MOKE, the sample presents an out-of-plane magnetization. (b) longitudinal MOKE, the sample has an in-plane magnetization, which is parallel to the plane of incidence. (c) transverse MOKE, the magnetization of the sample is in-plane, but perpendicular to the plane of incidence. The directions of the magnetization are indicated by black arrows.

MOKE is widely used to investigate the magnetic property of various materials. A longitudinal MOKE is used to investigate the magnetic property of Fe-phen/LSMO systems in Section 6.3. The detail of the experimental set-up is given in Section 3.1.2.

3 Experimental methods

Scanning tunneling microscope (STM) has been widely spread and rapidly developed since it was invented in 1981. Without doubt, STM is a powerful tool for research in condensed matter physics, especially in the field of surface science, because it can investigate and manipulate the surface at molecular or even atomic level under particular conditions. In this dissertation, single magnetic molecules on surfaces were mainly investigated with STM, which is obviously surface sensitive. Therefore, the samples investigated with STM require very high quality. In this chapter, firstly the general experimental facilities mainly involved will be introduced. Subsequently, sample preparation processes will be described. Eventually, the details of the STM techniques used in the experiment will be explained.

3.1 Experiment facilities

The main experimental measurements in this dissertation were performed on three individual home-built cryogenic STM facilities. All of them share a similar construction, which consists not only an STM chamber but also a preparation chamber and a load-lock chamber. These three chambers share a rotary pump but they are independently pumped to ultra-high vacuum (UHV) by discrete molecular turbo pumps and ion getter pumps. Obviously, the three chambers are connected, thus objects can be moved from one chamber to another through valves by transfer rods without breaking UHV. A sketch about the main structure of preparation and STM chambers of the JT-STM is presented in Figure 3.1, but the load-lock chamber behind the preparation chamber is out of sight. The details about each chamber will be depicted later in this section. As a whole set-up, three chambers together with other accessories fixed on a supporting frame. The whole set-up can be lifted up by four pneumatic vibration isolator feet (Newport S-2000 series). Therefore the STM measurement is isolated from the vibrations transmitted from the ground.

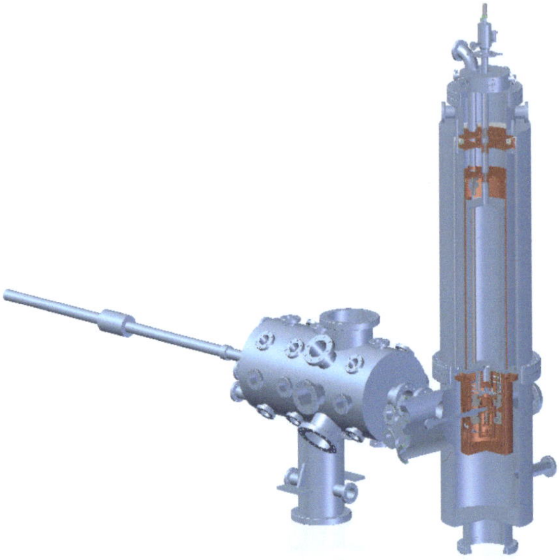

Figure 3.1: A skectch about the JT-STM without pumps. The load-lock chamber behind the preparation chamber is out of sight.

3.1.1 STM chamber

All STM chambers are equipped with a liquid helium cryostat encircled and cooled by a liquid nitrogen cryostat, which offers a base temperature of 4.2 K for the first STM, which is named as 4 K STM. To achieve a lower temperature, a Joule-Thomson expansion stage is supplemented in the second STM, called JT-STM [35]. When pure ^{4}He gas was used as medium, the lowest available temperature is 930 mK. A ^{3}He/^{4}He mixture is used in this dissertation, which guarantees a stable temperature of 800 mK under continues mode, after stopping condensing, 650 mK can be kept for hours before all of the condensed mixture consumes. On the third STM, a dilution unit rather than a Joule-Thomson stage is installed, therefore a temperature as low as 20 mK on the STM head can be realized. This STM is called Dilution-STM.

All STM heads reside in a magnetic coil made by superconducting NbTi wire. The maximum out of plain fields can be applied on the samples are 250 mT, 3 T and 6 T for 4 K STM, JT-STM and Dilution-STM, respectively.

The base pressures in all STM chambers persist in the low 10^{-10} mbar range under the work of ion getter pumps. Actually, the cryostat acts like cryogenic pump or sorption pump that also gives a contribution to the good pressure. Additionally, a non-evaporable getter pump purchased from *SAES* is installed on each STM chamber to absorb hydrogen and other active gases which constitute the main part of residual gas in addition to traditional UHV pumps.

3.1.2 Preparation chamber

The preparation chamber offers a space to flash and sputter tungsten tip, anneal and sputter crystals (see next section). Depositions of metallic materials from evaporators are also carried out in these chambers. The base pressure is kept below 1×10^{-11} mbar. A differentially pumped ion sputtering gun, a manipulator with heating filament, various deposition sources (evaporators) for molecular beam epitaxy, a low-energy electron diffraction (LEED), a mass spectrometer and other accessories about UHV technique are mounted on the preparation chambers.

Longitudinal MOKE set-up

Additionally, a magneto-optic Kerr effect (MOKE) (see Section 2.8) set-up is installed on the preparation chamber of the 4 K STM, which is used to investigate the magnetic property of Fe-phen/LSMO samples (see Section 6.3). In our experiment, the longitudinal geometry is used. A red laser source passing through a polarizer filter is employed as the linearly polarized incident light. After being reflected by the sample mounted on the manipulator, the out-coming light passes through an analyser polarizing filter thence reaches the detector, which contains a phototransistor. The change of the intensity of the reflected light is converted into change of the voltage signal, which is recorded to indicate the change of the relative strength of the longitudinal magnetization. The in-plane magnetic field is generated by an electromagnet. The external magnetic field applied to the sample can be changed by alternating the current in the coil, thus hysteresis loops can be swept. The temperature of the sample under measurement can be changed by cooling the head of the manipulator.

3.1.3 Load-lock chamber

To avoid venting the STM chamber or preparation chamber frequently, a small load-lock chamber is used as auxiliary chamber to load or withdraw objects such as STM tips, crystals and other components on sample plates from or to atmosphere directly, since it can be vented and pumped conveniently. Moreover, it can also differentially pump the lens region of the ion sputtering gun on preparation chamber. After pumping with molecular turbo pumps, the pressure in the load-lock chambers stay in the range of low 10^{-8} mbar, which is safe for opening the load-lock chambers to the preparation chambers in order to transfer objects. In this dissertation, molecular depositions (see next section) were also performed in load-lock chamber rather than preparation chamber to prevent contamination.

3.2 Sample preparation

As one electrode of the tunneling junction of STM, the substrate or the sample must be a conductor, or at least semiconductor with significant conductance. In this dissertation, we use various single-crystal metals with different crystal orientations, e.g. Cu(100), Cu(111), Ag(111) & Au(111), and thin films of lanthanum strontium manganite (LSMO) regarded as half-metal on strontium titanate (STO)(100).

In order to study molecules at nanometre scale, the metallic substrates (including Cu, Ag and Au) must be prepared in following process to achieve an atomically flat and clean surface. At first, the top-most layers on the crystals containing contaminates are completely removed by sputtering the surfaces with Ar^+ ions at a kinetic energy of 3 kV under a pressure about 3×10^{-7} mbar for approximately one hour. Subsequently, the crystals are annealed to a temperature 450 °C under a pressure better than 1×10^{-9} mbar to allow the atoms to diffuse in order to form an atomically flat surface. The temperature is measured by an infrared pyrometer. After above two treatments, the crystals should be ready to deposit molecules, otherwise a higher quality can be achieved by simply repeating the process several times. Frequently, molecules are not deposited on pure metallic crystal surface directly. Some complicated interfaces are required to realize particular functions. In the following, the processes to obtain single monolayer Cu_2N and Fe_2N on Cu(100) surface, and metallic thin film grown on metallic crystal (e.g. Co on Cu(100)) will be introduced.

The single monolayer of Cu_2N on Cu(100) [36] can be simply obtained by bombarding the well-prepared Cu(100) with surface with nitrogen ions at a kinetic energy of 650 V under a pressure about 3×10^{-8} mbar for approximately one hour followed by annealing the crystal to 300 °C [37]. The process to obtain single monolayer of Fe_2N on Cu(100) [38] is almost same as that of Cu_2N except one additional step should be inserted. Iron has to be deposited onto the surface after bombarding with nitrogen ion and before post-annealing. Since part of the iron atoms are embedded into the copper bulk during post-annealing, approximately two monolayers of iron are required to fabricate one monolayer of Fe_2N on Cu(100). As the process to prepare Fe_2N/Cu(100) is longer and more complicated compared to pure metal surfaces and Cu_2N/Cu(100), the surface of Fe_2N/Cu(100) contains more contaminates (see Figure 3.2). With respect to the remaining atomically flat areas, the quality of Fe_2N is sufficient for the study of single molecules.

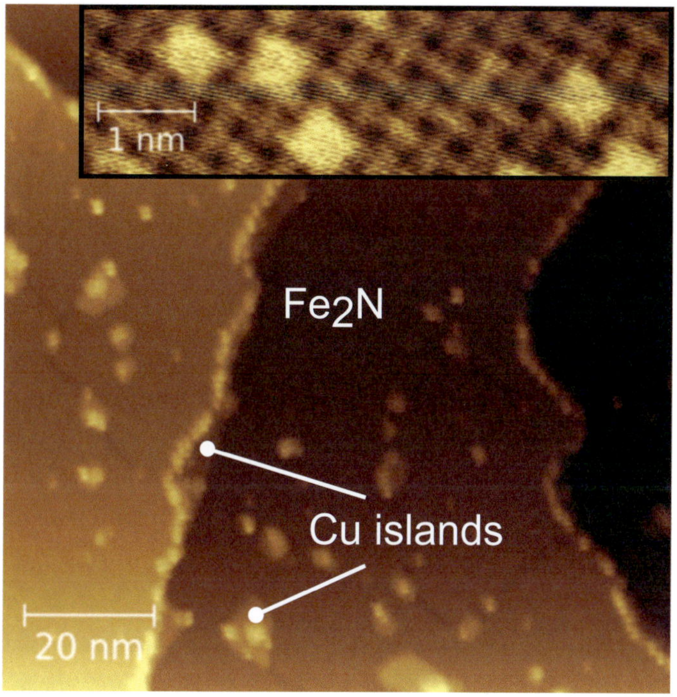

Figure 3.2: An STM image of Fe_2N/Cu(100). Most of the surface is covered by Fe_2N layer, except some Cu islands. The inset shows an atomically resolved image. One can recognize the specific reconstruction of Fe_2N.

Iron is sublimed from an evaporator by heating the tip of a pure iron rod (99.99%) with an electron beam at power of 1 kV× 10 mA under a pressure better than 5×10^{-10} mbar [39]. The calibrated deposition ratio is 0.6 ML/min. Correspondingly, cobalt can also be sublimed by a power of 560 V× 16 mA. The estimated deposition rate is 0.1 ML/min. In this way, after a post-annealing to 100 °C, cobalt thin films were grown on Cu(100), which was used to deposit manganese phthalocyanine in this dissertation.

The only non-metallic substrate we used is LSMO/STO(100). It is fabricated by the group of Prof. Alex Dediu in Bologna, Italy [40]. The thickness of the LSMO layer is between 6 and 10 nm. Since annealing LSMO to a high temperature in UHV will induce oxygen vacancies, the LSMO/STO(100) substrates were only heated to 140 °C. Obviously, sputtering LSMO/STO(100) is avoided. According to our knowledge, atomic resolution on LSMO/STO(100) exposed to air by STM has not been achieved. The quality of the surface of LSMO/STO(100) after preparation is shown in Figure 3.3. Steps of unit cell height (about 0.4 nm) are visible in line profile.

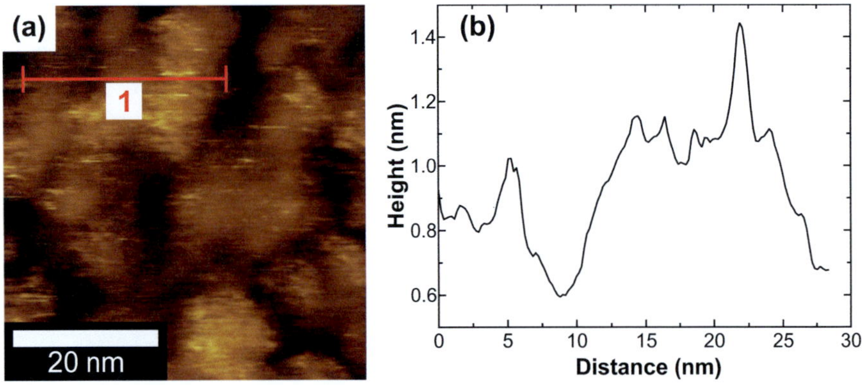

Figure 3.3: (a) An STM image of LSMO/STO(100) after preparation, set point: 1 V, 0.5 nA. (b) Line profile along the red line labeled by "1" in (a).

After the substrate is prepared, the various molecules were deposited onto room-temperature substrates under a pressure better than 1×10^{-7} mbar by subliming molecular powder from a crucible. In this dissertation the sublimation temperature varies a lot for different kinds of molecules, e.g., 30 °C for M-bis(2,2,6,6-tetramethyl-3, 5-heptanedionato) (M = Co, Ni and Cu) (see Section 4.1, 380 °C for manganese phthalocyanine (see Section 4.2), 62 °C for tris(2,2,6,6-tetramethyl-3,5-heptanedionato)terbium(III) (see Section 5.1), 160 °C for $[Tb(tmhd)_3]_2$bpm (see Section 5.2) and 170 °C for Fe(1,10-phenanthroline)$_2$(NCS)$_2$ (Fe-phen) (see Chapter 6). The deposition rate was about 0.1 monolayer/s. Sub-monolayer of molecules were deposited by exposing the substrate to the molecular beam for several seconds after degassing the molecules at a temperature slightly below the sublimation temperature for hours.

The samples were transferred into the STM immediately after preparation to start cooling down. To decrease the amount of contaminates on the surface, during the whole evolution of sample preparation, from treating the substrate to transferring sample to STM, UHV was preserved. Furthermore, the time of preparation was reduced as short as possible.

3.3 Scanning tunneling microscope techniques

The first scanning tunneling microscope was invented by Gerd Binnig and Heinrich Rohrer in IBM Zürich in 1981 [41, 42]. Because of this significant contribution, those two inventors earned a quarter of the Nobel Prize in Physics 1986 respectively by sharing the rest half with Ernst Ruska, who established the fundamentals in electron optics and designed the first electron microscope. After the prototype, scanning tunneling microscopy was further developed and improved rapidly in 1980s. As the first technique to achieve atomic resolution in real space, STM can be used in a broad temperature range from millikelvin to above 1000 K in various conditions, e.g. not only in UHV, but also in air, water and other liquids [13]. Since its incomparable advantages, STM has been wildly applied in surface science, nowadays.

In this section, the information concerning scanning tunneling microscope from basic concepts to advanced techniques will be introduced.

3.3.1 Basic concepts to operate STM

Based on the concept of quantum tunnneling (see Section 2.5), electrons under a bias can tunnel through the junction between conducting tip and sample when the tip is sufficiently close to the sample. Generally speaking, STM is an instrument to record the tunneling current at different modes in various conditions as a function of location, tip-sample distance, bias, temperature, magnetic field and other external stimuli.

As depicted in Figure 3.4, an STM set-up consists a sample in a sample rack, a scanning head, electronics to control the scan and to deal with the tunneling current, and a computer to process and display data. The components of the scanning head include a piezoelectric tube with a scanning tip at the end, scanner-holder to fix the tube and piezos for coarse motion of the whole scanner-holder.

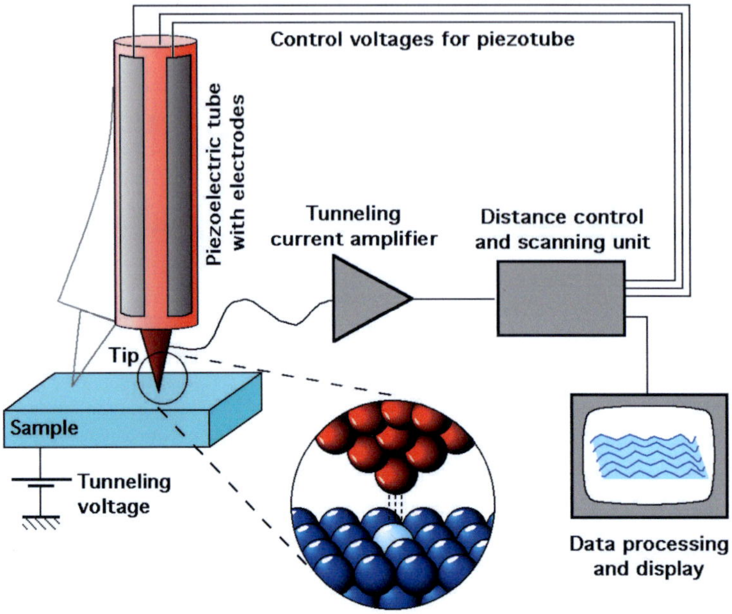

Figure 3.4: A sketch inllustrates the basic concepts of an STM. This figure is attributed to Michael Schmid, TU Wien.

Tip preparation

As a challenge, besides clean and flat samples (see Section 3.2), high-resolution STM also strongly depends on high-quality tips, which means a sharp and stable apex. Usually, tips can be made either electrochemically (etched tungsten wire) or mechanically (sheared platinum-iridium or gold wire). To obtain tip with atomic scale apex or with particular functions, more complicated treatments were developed, for instance, tips with attached carbon nanotubes or milled tips using focused-ion beam have been used [43, 44]. Tungsten tips coated by a magnetic layer (e.g. cobalt, iron or nickel) are used for spin-polarized STM.

In this dissertation, all of the tips were prepared from tungsten wire by chemical etching in a solution of sodium hydroxide (e.g. NaOH). After flashing the apex of the raw tip above the melting temperature in UHV, the oxidized layer on the surface can be sublimed, moreover, carbon and other impurities in the wire precipitate to the surface, which can be removed by subsequent Ar^+ sputtering. The final post-flashing makes the apex sharp and smooth.

Motion

To realize convenient operation in the experiment, tips and samples must be able to be moved relatively not only in nanometre scale but also in millimetre scale. In our STM, the whole sample rack can be moved laterally by shear-piezos in 5 mm range, similarly, the STM scanner-holder can move vertically. This quick and long range movement is called coarse motion. After bringing the tip close to the surface by vertical coarse motion, the three-dimensional fine motion of tip on a small area in the micrometre range of the surface is controlled by the piezoelectric tube, which stays in the scanner and holds the tip at the end. The typical distances between the tip and sample are in the range from 4 Å to 7 Å, which should sit at the equilibrium position between attractive and repulsive interactions [13].

Two modes

During scanning across the surface, the tip stops at periodic positions to measure the tunneling current as main information to examine the surface. As discussed in Section 2.5, the tunneling current strongly depends on the tip-sample distance and the density of states on the measuring point. Usually, the measuring points constitute an array, the measured information is saved as a matrix, which can be illustrated in a map, each pixel refer to a corresponding measuring point.

Since the tunneling current exponentially depends on the tip-sample distance, by keeping the height of the tip in z direction during scanning, the tunneling current can precisely monitor the topography of the surface. This scanning way is called constant-height mode. In an atomically flat area, under the constant-height mode, the density of states on the surface can thus be monitored. Besides the constant-height mode, another mode called constant-current mode is used more frequently. In this mode, the tunneling current is preserved by regulating the height of the tip via an electronic feedback loop to adjust the voltage applied on the piezoelectric tube. Actually, in constant-current mode, the movement of the tip in z direction refers the fluctuation of height and density of states on the surface directly. Therefore, the topography of the surface can be depicted by the trace of the tip in z direction, which contributes the contrast on the displayed image. Obviously, the topographic image show mostly the geometrical information when the density of states of the conducting sample is homogeneous on the surface. Thus, only the heights of step of terrace, reconstruction or other crystal defects on the surface can be declared by constant-current image. Comparing to constant-height mode, constant-current mode is safer against crashing the tip into the obstacles on rough surfaces. However, the scanning speed is slower, since the feedback loop and piezoelectric tube require time to adjust the height of the tip.

Resolution

To achieve atomic resolution, it is considered necessary to have a 1 Å lateral resolution and a 10 pm z resolution in space.

Besides the spacial resolution, energy resolution is also critical for tunneling spectroscopy, i.e. to observe different energy levels of states, inelastic excitations (see Section 2.6.2), occupied and unoccupied orbitals. As discussed in last chapter, an unavoidable limit for energy resolution is the thermal broadening of the electron energy distribution on the Fermi level at finite temperatures. As we know, the smearing in energy at a particular temperature is scaled by $k_B T$, which is approximately 0.026 eV at room temperature. Considering the thermal broadening on both the tip and sample, the energetic smearing of an STM measurement is already as large as 50 meV at room temperarute, while it is only 0.7 meV at 4.2 K. Consequently, many STM measurements are operated at low temperatures, not only to avoid the lateral motion of adsorbate on the surface, but also to obtain a better energy resolution which is linear to temperature. The energy resolution of real measurement also depends on the precision of original electronic measuring signal. Especially at dozens of millikelvin, the thermal broadening is below 10 μeV, and a precision of 10 μV is necessary in the bias voltage and the voltage noise of the current detection circuit.

Noise control

To achieve a high resolution, the STM must be isolated from mechanical vibration, which is guaranteed by an air-floated frame (see Section 3.1) where the whole STM chamber sits on. Furthermore, by using three springs, the STM head is hung below the cryostat to damp the vibrations from the chamber.

In parallel, irregular electrical noise also disturbs the STM measurement. Thus, all of the cables and devices for measuring must be well screened or grounded. All electrical power of facilities not used during the STM measurement should be cut.

A small magnetic field (in the range of hundreds of mT) can be applied around the sample and tip to damp vibrations due to adding currents and thus to reduce the noise.

Sophisticated electronics

As shown in Figure 3.4, the STM facility requires a sophisticated electronic system. We use commercial control-system and software from Nanonis to control the both coarse and fine motions of the STM. The tunneling current in the nanoampere range is amplified to several voltages by an *I-V* converter. All data are processed on computers equipped with various softwares. For instance, the original images produced by STM are saved as matrices. After data acquisition, an optional colour code can be used to demonstrate the features on the image visually. The contrast and clarity can also be improved by following treatment. Additional dual phase analog lock-in amplifiers (Princeton Applied Research/Ametek SIGNAL RECOVERY 5210) are used in scanning tunneling spectroscopy, which will be introduced in detail in next subsection.

3.3.2 Measuring scanning tunneling spectra

In comparison to scanning over the surface at a fixed bias to depict images, the STM tip can also hover at a particular position to scan a spectrum by sweeping the bias voltage. This measurement mode is called scanning tunneling spectroscopy (STS). In this mode, usually, the dependences of tunneling current (*I*), first derivative of current, i.e. differential conductance (dI/dV), as well as the second derivative of current (d^2I/dV^2) on bias are recorded and paid close attention. Since the differential conductance refers the density of states, STS is a powerful method to investigate the local electronic properties of the sample as a function of energy (corresponding to the bias in the measurement). As an extremely local measurement, STS can resolve the density of states on the surface with atomic resolution. The contrast in the density of states at a particular energy can be presented in a map of the dI/dV signal, which usually graphically reveals the spacial distribution of orbital in this dissertation. Moreover, it can also be used to indicate the contrast of chemical composition, magnetic structure, surface states and additional adsorbate on the surface. Experimentally, as a derivative, STS can be obtained either numerically from the $I(V)$ curves or by using a lock-in amplifier.

Numerical derivation

When the measured data points of an $I(V)$ plot is sufficient (i.e. ΔV is small enough), the first derivative can be approximately represented by $\Delta I/\Delta V$, the ΔI and ΔV refer to the differences of I and V between neighbouring measured points. Hence the second derivative can be calculated from the $\Delta I/\Delta(V)$ in same way. This treatment is called numerical derivation. Obviously, a smaller ΔV results in more accurate numerical derivation in theory. However, the actual precision is also limited by the capability of electronics to measure a small current change (ΔI) precisely after a small bias change (ΔV).

Analog lock-in technique

In comparison to numerical derivation, usually analog lock-in amplifiers give a more precise STS signal with a small modulation voltage. In other words, lock-in technique has a higher resolution for STS. The modulation voltage refers to an alternating signal, $\Delta V\cos(\omega t)$, added to modulate the DC bias V_0. Here, ΔV is the modulation amplitude. This modulation voltage results in AC components in the tunneling current. As a function of bias, the tunneling current can be represented by Taylor expansion at the point of V_0:

$$\begin{aligned} I(V(t)) =& I(V_0) + \frac{1}{1!}\frac{dI}{dV}(V(t)-V_0) + \frac{1}{2!}\frac{d^2I}{dV^2}(V(t)-V_0)^2 \\ &+ \frac{1}{3!}\frac{d^3I}{dV^3}(V(t)-V_0)^3 + ... \end{aligned} \tag{3.1}$$

Since the bias is a mixture of DC and AC components $V(t) = V_0 + \Delta V\cos(\omega t)$, we find:

$$\begin{aligned} I(V(t)) =& I(V_0) + \frac{1}{1!}\frac{dI}{dV}(\Delta V\cos(\omega t)) + \frac{1}{2!}\frac{d^2I}{dV^2}(\Delta V\cos(\omega t))^2 \\ &+ \frac{1}{3!}\frac{d^3I}{dV^3}(\Delta V\cos(\omega t))^3 + ... \end{aligned} \tag{3.2}$$

Now the aim of the lock-in technique is to extract the first derivative of $I(V)$, i.e. dI/dV. To realize that, the lock-in electronics multiplies the tunneling current with phase corrected modulation voltage $\Delta V\cos(\omega t + \phi_0)$, then averages the product in a sufficient time range:

$$\begin{aligned} &\frac{1}{T}\int_{t-T}^{t} I(V(t'))\Delta V\cos(\omega t' + \phi_0)\,dt' \\ =&\frac{1}{T}\int_{t-T}^{t} I(V_0)\Delta V\cos(\omega t' + \phi_0)\,dt' \\ &+\frac{1}{T}\int_{t-T}^{t} \frac{1}{1!}\frac{dI}{dV}\Delta V^2(\cos(\phi_0) + \cos(2\omega t' + \phi_0))\,dt' \\ &+\frac{1}{T}\int_{t-T}^{t} \frac{1}{2!}\frac{d^2I}{dV^2}\Delta V^3(\cos(\omega t'))^2\cos(\omega t' + \phi_0)\,dt' \end{aligned} \tag{3.3}$$

$$+\frac{1}{T}\int_{t-T}^{t}\frac{1}{3!}\frac{d^3I}{dV^3}\Delta V^4(\cos(\omega t'))^3\cos(\omega t'+\phi_0)\,dt'+...$$

Obviously, after a long enough integration time T, only the DC component is kept in the second therm, which is proportional to dI/dV, when a proper ϕ_0 is chosen, while the other oscillating terms vanish. Therefore, the output of lock-in signal can be regarded as dI/dV, i.e. the differential conductance at bias of V_0. Consequently, by averaging the product of the tunneling current multiplied by the modulation voltage, the differential conductance can be extracted. Similarly, the derivative of order n can be extracted by multiplying the tunneling current with frequency-multiplied modulation signal $\cos(n\omega t+\phi)$. Among them, the second derivative is frequently detected.

In the experiment, a proper frequency of modulation, $(\omega/2\pi)$, must be chosen to avoid any undesired resonance. Additionally, in order to scan the STS at various bias quickly, a higher frequency is preferred to save integration time, which should be at least a couple of periods of modulation. However, the frequency is limited by the bandwidth of the *I*-*V* converter, which depends on the gain. A commonly used frequency is this dissertation ranges from hundreds to thousands Hertz. To increase the magnitude of the lock-in signal, besides choosing an optimal phase shift ϕ, another way is to increase the amplitude of modulation voltage ΔV. However, a larger ΔV not only harms the energy resolution (according to Equation 2.24), but also induces larger noise originating from the high orders in Equation 3.3. Therefore, the optimal ΔV is determined by the signal-noise ratio and the energy resolution, which are limited by the capacity of electronics and the measuring temperature as discussed in last section. Under ideal situations, to achieve sufficient lock-in signal while keep the high energy resolution, ΔV should be set comparable to $k_B T/e$. Meanwhile, the difference of bias between neighbouring points on STS can be half of ΔV. Of course, this critical condition requires an extremely low background noise for the whole facility, especially at low temperatures.

4 Magnetic molecules with single 3*d* ions: $M(thd)_2$ and MnPc

In the field of single magnetic molecules (SMMs), the magnetic moment is usually contributed by transition metal ions as the cores of the molecules. Since 3*d* ions are better known to scientists and complexes containing them are easier and cheaper to synthesize compared to other transition metal elements, complexes containing 3*d* ions have played an important role since the primary time of SMM. Actually, a great advantage of 3*d* ions is that their orbitals and spins are usually strongly coupled to their environments. Thus, the spin states and subsequent magnetic properties can be dramatically changed by manipulating the ligand field.

Because of the absence of magnetic interactions between core ions, SMMs with single ion, some times referred as single-ion magnets (SIMs) in literature, are regarded as the simple examples to learn about the splitting of the orbitals and hence the breaking of the degenerate spin states of the central ion. Regarding 3*d* SIMs, the first example was $K[(tpa^{Mes})Fe]$ reported in 2010 [45], which presents a high-spin ground state. Afterwards, as the very common magnetic 3*d* elements in the nature, Fe, Co, Ni, Mn and Cr have been widely use to synthesize SIMs [46]. However, the magnetic properties of these compounds were evidenced by measuring the magnetic susceptibility of crystals.

Thanks to transport measurements in break junctions and the junctions of scanning tunneling microscopy (STM), it became possible to detect the magnetic properties of a single molecule between two non-magnetic electrodes [47]. Particularly for STM, SIMs are locally investigated with the precision of sub-nanometre due to STM's high spacial resolution. In the pioneering works of A. Heinrich et al. [48, 49], it was shown that magnetic excitations of the ion from its ground state or ground state doublet can be observed with inelastic tunneling spectroscopy. The excitations show up as symmetric steps in the differential tunneling conductance dI/dV marking the energy of the excitation from the lower energy state to an excited state via spin-flip scattering of tunneling electrons [50, 49, 51, 52, 53]. By studying the energy dependence of the excitations as function of an external magnetic field, the spin Hamiltonian can be determined. Spin excitations in metal-organic molecules as function of the tip-molecule distance have recently been performed [54, 3]. A continuous and slight change in excitation energies was observed and was interpreted by a tip-molecule interaction. As in an STM, an electric current is needed, the magnetic ion is necessarily in contact with an electron bath, i.e. the substrate. This electron bath can also promote spin scattering with the local ion leading in first or higher orders to the screening of the localized spin by the Kondo effect (see Section 2.7). The presence or absence of Kondo effect is intimately linked to the nature of the spin Hamiltonian [23, 55, 56, 27, 57, 58]. Similarly, the Kondo effect was also observed in magnetic molecules in break junctions [59, 60, 61, 62].

In this chapter, a project about the local investigation of $M(thd)_2$ molecules by STM will be illustrated as the first example to introduce the characteristic of the SMM with single $3d$ ion. Systematic experiments were performed on metal-chelate molecules with various total spin S (Co, Ni and Cu as metal ions). The magnetic anisotropies of molecules were manipulated by the formation of a molecular contact with the tip of a scanning tunneling microscope. The modifications of the anisotropy result from the change of the energetic order of the spin states upon contact formation due to strong mechanical deformation of the rather flexible organic molecules. The spin states of the magnetic ions in the molecules are represented by the effective spin Hamiltonian describing the zero field splitting (ZFS) of the magnetic states as introduced in Section 2.2 and 2.3. We determined the ZFS by analysing magnetic field dependent inelastic tunneling spectra taken at temperatures down to 30 mK. When moving the tip towards the molecule, the tunneling current abruptly jumps to higher values indicating the sudden deformation of the molecule bridging the tunneling junction. Remarkably, with the formation of the contact, an abrupt change of the ZFS occurs, which involves the energetic crossing of eigenstates of the spin system, i.e. the origin of the complete change of the magnetic anisotropy of the system. These results implies that ZFS expected in mechanical break junctions can drastically deviate from those of adsorbed molecules probed by other techniques. This sharp transition in ZFS potentially allows to prepare superposition quantum states [63, 64], a prerequisite for their use as quantum bits. Additionally, Kondo resonances were observed in these systems under particular situation.

In parallel, another example of the SMM with single $3d$ ion, manganese phthalocyanine molecules were deposited onto Cu(100), and Co/Cu(100) and investigated with STM and STS, respectively. The adsorption geometries of the molecules extracted from topographies reveal how different monolayers stack. The Kondo resonance of MnPc on Cu(100) and spin-flip inelastic excitations reveal the magnetic properties of Mn ions.

4.1 M(thd)$_2$ molecules

To design an eye catching SIM, not only a proper central ion should be chosen, but also favourable ligands should be considered in order to split the orbitals appropriately. In this project, we were attracted by the magnetic properties of a series of metal-chelate molecules bis(2,2,6,6-tetramethyl-3, 5-heptanedionato) 3d metal(II) molecules (M(thd)$_2$) with different central 3*d* ions (Co, Ni and Cu) [65]. As displayed by the model in Figure 4.1(a), the central metal ion of an M(thd)$_2$ is surrounded by two thd groups, which are only loosely bound by chelate coordinate bonds. The thd groups themselves contain mainly sp^3 bonds that in principle allow the rotation around the bond axes. Thus, we expect that the molecule is rather flexible, i.e. can react on external forces by deformation as in contrast to fully conjugated molecules, which consequently should result in the modification of ZFS and hence the change in magnetic properties. By replacing the central ion among Co, Ni and Cu, the total spin S can be altered. Overall, the M(thd)$_2$ molecules offer us a platform to study the related spin states of the central ions under various configurations.

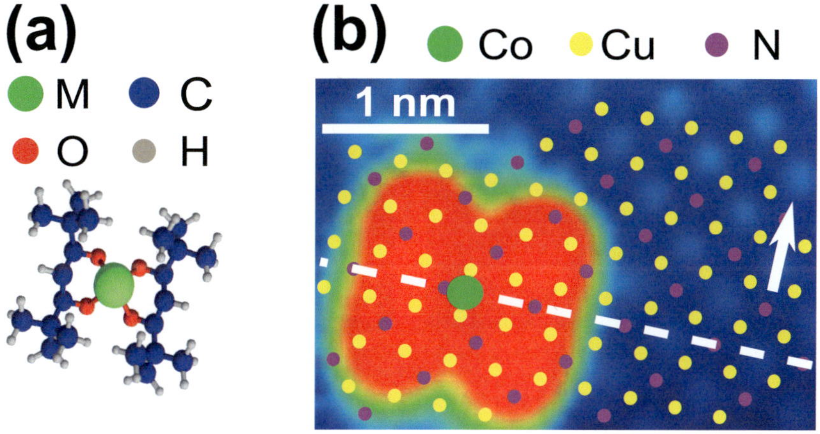

Figure 4.1: (a) Model of an M(thd)$_2$ molecule. (b) STM image of a Co(thd)$_2$ molecule adsorbed on atomically resolved Cu_2N/Cu(100) combined with a model of the Cu_2N layer. Dots with different colours refer to the positions of different atoms. The white arrow indicates the [110] direction of Cu(100) and the white dashed line the mirror plane of symmetry of the Co ion's surrounding. The image was scanned at 900 mK with set point: 100 mV, 40 pA.

4.1.1 Topography of $M(thd)_2$ molecules deposited on Cu_2N

The $M(thd)_2$ molecules were deposited onto $Cu_2N/Cu(100)$ substrates by subliming commercial molecular powder at 30 °C from a crucible (The detail of sample preparation has been introduced in Section 3.2). The STM image of a $Co(thd)_2$ molecule on an atomically resolved $Cu_2N/Cu(100)$ surface reveals the adsorption site of the molecule in the force fields of the surface (see Figure 4.1(b)). Note that the central Co ion is surrounded by two symmetric thd groups. However, only a mirror symmetry with respect to the Cu_2N substrate is present. Thus, the symmetry of the free molecule is reduced on the surface to a mirror operation indicated by the white dashed line. The system therefore has a C_{1v} symmetry, which is confirmed by the differential conductance dI/dV map on a single $Co(thd)_2$ molecule (see Figure 4.2(b))in stark contrast to the topography of the same molecule (see Figure 4.2(a)) presenting a four-fold symmetry (see Figure 4.2(a)). Identical experiments reveal that the C_{1v} symmetry fit $Ni(thd)_2$ and $Cu(thd)_2$ molecules adsorbed on Cu_2N.

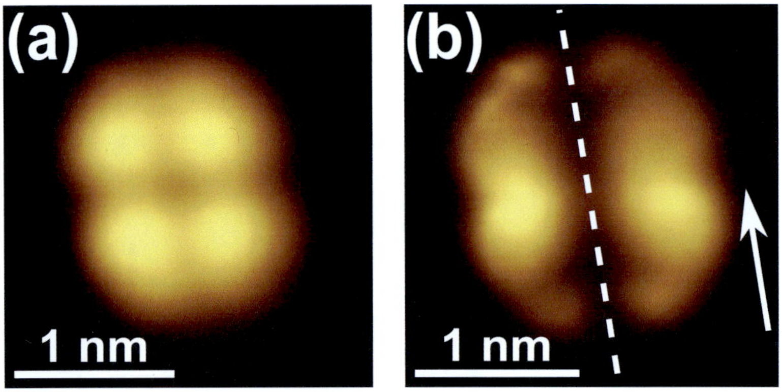

Figure 4.2: Topography of a $Co(thd)_2$ molecule (a) presents a four-fold symmetry, while the corresponding map of dI/dV signal on same molecule (b) shows a C_{1v} symmetry with a mirror plane (white dashed line). The white arrow indicates the [110] direction of Cu(100) surface. Temperature: 900 mK, set point: 720 mV, 400 pA, modulation of lock-in: 10 mV RMS.

4.1.2 Molecular deformation due to contact formation

As the molecule is rather flexible, we expect that the interaction between the tip and the molecule can lead to changes of the adsorption geometry [58]. To check for these effects, we measured $I(z)$ curves by recording the tunneling current I as function of the tip position

z while approaching the tip towards the molecule. In the experiment, the tip was first placed laterally over the center of the molecule above the Co ion under tunneling conditions representing a large tip-sample distance, i.e. at a low current of 20 pA under a bias voltage of 20 mV. Then, the feedback loop was halted and the tip was slowly approached towards the substrate by a z-offset while measuring the current.

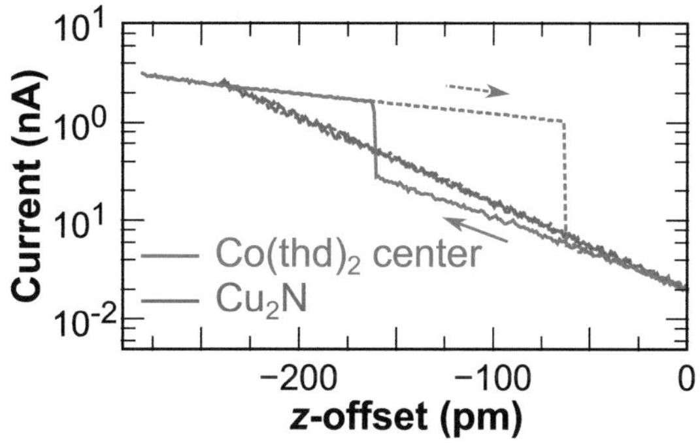

Figure 4.3: $I(z)$ curves on Co(thd)$_2$ molecule center (green) and Cu$_2$N (gray). Negative z- offsets represent a decrease of the tip-sample distance from the initial set point: 20 mV, 20 pA. The arrows indicate the direction, approaching and retracting the tip. Note that the current is in a logarithmic scale.

The $I(z)$ curve shown in Figure 4.3 were measured by ramping the z-offset in a loop from 0 pm to -280 pm and backwards while keeping the sample bias constant, in other words, approaching the tip 280 pm from the initial set point and retracting tip subsequently. In stark contrast to the trivial exponential rise of the current observed on the bare Cu$_2$N surface (gray line), the tunneling current through the Co(thd)$_2$ molecule displays a prominent jump at an offset of about -160 pm from 0.18×10^{-3} G_0 to 1.0×10^{-3} G_0, where $G_0 = \frac{2e^2}{h} \approx 7.7 \times 10^{-5}$ S is the quantum of conductance, followed by a relatively slow increase upon further approach (green solid line). When retracting the tip again, a hysteresis was observed (green dashed line) and the current returns to the same values at an offset of about -60 pm. The hysteresis indicates a closing and opening of a molecular contact involving molecular deformation [66, 67, 68, 69, 70]. Similar hysteresis in closing and opening curves are commonly observed in break junctions with molecules in between [71, 72, 73]. Accordingly, we label the two states as tunneling and contact regime respectively. Images after contacting show no modification of the molecule or its adsorption site. This indicates that the molecule is only slightly deformed during contact returning to its original state after breaking the contact. These observations also exclude chemical modifications due to the process of closing and opening again the contact.

Besides on Co(thd)$_2$, a similar behaviour with a jump to contact upon approaching the tip was also observed on Ni(thd)$_2$ and Cu(thd)$_2$, despite the slight shift of the positions of the jumps (see Figure 4.4). Thus, these are general properties of the organic ligands and are not affected by the central ion confirming the interpretation of the molecular distortion by the presence of the tip.

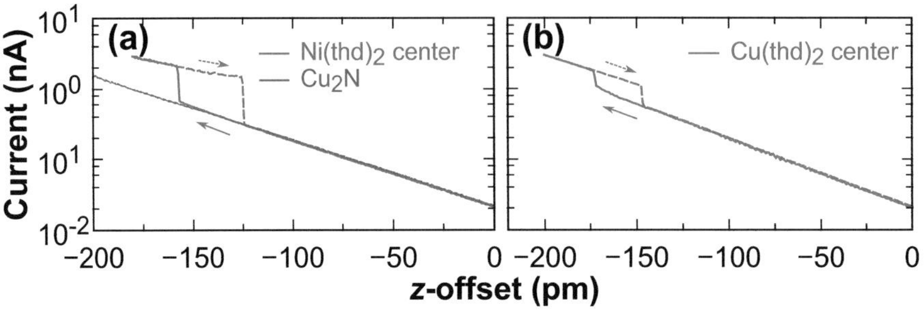

Figure 4.4: $I(z)$ curves on Ni(thd)$_2$ (a) and Cu(thd)$_2$ (b) molecules. The tip starts to approach from the same initial set point: 20 mV, 20 pA.

4.1.3 Inelastic excitation and Kondo resonance on Co(thd)$_2$

To study the magnetic behaviour of the molecule by using a lock-in technique (see Section 3.3.2), we recorded dI/dV spectra on Co(thd)$_2$ in the tunneling regime (at 200 pA current, i.e. before the jump to contact) and in the contact regime (at 2.5 nA, i.e. after contact formation). These two conditions are characterized by two different dI/dV spectra (see Figure 4.5). The spectra display symmetric steps below 10 mV indicative of an inelastic excitation. Note that the spectra were individually normalized by dividing with the conductance at large bias, in order to reveal the magnitudes of inelastic excitations compared to the elastic tunneling background. As we will show below, the energies of the excitations depend on magnetic field confirming a magnetic origin [48, 74]. Additionally, a sharp resonance peak at the Fermi energy is observed for both conditions. This resonance can be split by a magnetic field confirming a Kondo resonance, as discussed below. In contrast to the spectra on molecules, data taken on bare Cu$_2$N (gray dots) show relatively flat spectra without clear signs of inelastic excitations or zero bias peaks. Clearly, in the contact condition (red dots) inelastic excitations move to higher energies than in the tunneling condition (blue dots). Thus, the ZFS of the magnetic states is modified by the formation of the molecular contact.

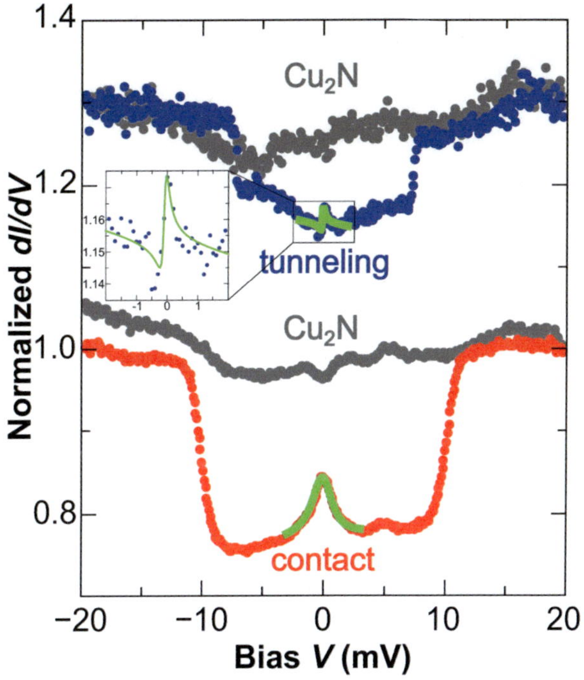

Figure 4.5: Normalized dI/dV spectra on $Co(thd)_2$ with set points of 20 mV, 200 pA (blue) and 20 mV, 2.5 nA (red), corresponding to z-offset of -140 pm and -250 pm, regarded as tunneling and contact conditions, respectively. The gray spectra were recorded on bare Cu_2N under identical tunneling conditions to the spectra on the molecules. The two groups of spectra are vertically offset by 0.3 for clarity. Green solid lines are fits of a Fano function. Spectra were measured at 900 mK with a modulation of 0.2 mV RMS at 487 Hz.

In parallel, the Kondo peak in contact is much more pronounced and wider than in tunneling. This can be explained by a stronger coupling between conduction electrons and the local spin of the Co ion due to the closer tip. As introduced in Section 2.7.3, the Fano function (green solid lines in Figure 4.5) was used to fit the Kondo resonances (the details about fitting are displayed in Table 4.1). Note that the Kondo temperature T_K extracted from fitting increased from 1.2 ± 0.5 K to 11.4 ± 0.2 K upon contact formation. Further note that the electronic coupling Γ of the localized spin to the electron bath largely determines the Kondo temperature. An increase of Γ results in first approximation to an exponential increase of T_K [75]. This makes the Kondo temperature a very sensitive indicator for the coupling. The Kondo temperature in the tunneling regime is similar to those reported for single Co atom on $Cu_2N/Cu(100)$ [27], but much lower than that of single Co atom on bare metal surfaces [76], since the interaction between Co ion and conduction electrons from copper substrate is decoupled by Cu_2N. In our case, the Co ion is decoupled from the substrate by both the Cu_2N and the organic ligands. In comparison, after molecular contact, the local spin on Co ion can additionally couple to conduction electrons from the tungsten tip, which results in a higher Kondo temperature and in a larger current through the junction.

Thus, in a short summary, when approaching the tip towards the Co(thd)$_2$ molecule, the abrupt increase in current indicates the point of contact formation with the molecule, in which the molecule suddenly deforms to bridge the gap between surface and tip. This results in modifications of the ZFS indicated by the change of the inelastic excitation energies and intensifies the coupling of the local spin to the electron bath increasing the Kondo temperature.

Table 4.1: Determined coefficients of fitting Kondo resonances near zero bias on dI/dV spectra of Co^{2+} in both tunneling and contact conditions in Figure 4.5 with Fano function: $y(x) = R_0(q + \frac{x - E_K}{k_B T_K})^2/(1 + (\frac{x - E_K}{k_B T_K})^2) + kx + y_0$, where T_K, q and E_K are the Kondo temperature, Fano parameter and position of Kondo resonance, respectively, $kx + y_0$ is a linear background.

	tunneling	contact
T_K (K)	1.2 ± 0.5	11.4 ± 0.2
q	-0.6 ± 0.2	0.07 ± 0.02
E_K (mV)	-0.08 ± 0.03	-0.01 ± 0.03
R_0	-0.021 ± 0.006	-0.0725 ± 0.0007
k	-0.0025 ± 0.0008	0.0019 ± 0.0004
y_0	1.174 ± 0.006	0.8438 ± 0.0006

4.1.4 Modification of ZFS due to molecular deformation

To investigate the details of the molecular deformation induced by the tip, dI/dV spectra on the center of Co(thd)$_2$ were recorded at various z-offsets from the initial set point (20 mV, 20 pA) as indicated in Figure 4.6(a). We here focus on the energy range, where the inelastic step is observed. Clearly, the spectra illustrate that changing the z-offset influences the position of the inelastic step. For z-offsets before the jump to contact (indicated by a blue ellipse) we observe a slight tendency of the excitation to shift to lower energies (position of the step is indicated by green dots). Under these tunneling conditions, the molecule has not yet bridged the contact but still lies on the Cu_2N surface, although the Co ion starts to feel the presence of the tip (c.f. Figure 4.6(b)). Thus we ascribe the change of the ZFS to modifications of the ligand field caused by the tip approaching the molecule. When, however, the molecular junction is suddenly formed, both the relative magnitude of the inelastic excitation step is significantly enhanced and the energy of the excitation abruptly increases. This illustrates that the ZFS of adsorbed molecules may largely differ from that of contacted molecules in break junctions, and shows that ZFS parameters of molecules in junctions cannot be taken from e.g. measurements of bulk probes. The further approach leads to a further increase of the excitation energy (indicated by a red ellipse). We ascribe the sudden modification of the ZFS to the jump of the molecule from tunneling into the contact configuration. The hysteresis in the contacting curve indicates that the two configurations are local minima in the configuration space and the jump corresponds to a quick transition of the molecule between the two states. Similarly, further deformation of the molecule leads to additional changes of the ZFS, but without jumps. Here, the configuration of the molecule is changed continuously with approaching the tip. We stress that particularly the sudden distorts of the molecule during contact formation, which is induced by interaction between tip and molecule, abruptly modify the ligand field acting on Co ion (see Figure 4.6(c)). Finally, the inelastic excitation energy starts to decrease again after the z-offset exceeds $-220\,\mathrm{pm}$ and returns to values similar to the tunneling condition at about $-300\,\mathrm{pm}$ (indicated by pink ellipse). This can be explained by the approaching tip pressing the molecule flat again (see Figure 4.6(d)).

Note that although the observed features in the tunneling spectra are moved in energy by contact formation, the qualitative shape of the spectra, e.g. the number of inelastic steps or the existence of a Kondo resonance, is not altered. This excludes charge transfer processes and changes of the spin of the magnetic ion due to contact formation.

4.1.5 Field dependent spectra on $Co(thd)_2$, $Ni(thd)_2$ and $Cu(thd)_2$

To study the ZFS in more detail, experiment operation was transferred from the JT-STM to the Dilution-STM in order to achieve a magnetic field 6 T and simultaneously a temperature as low as 30 mK (see Section 3.1). dI/dV spectra of all three molecules were recorded both in the tunneling and the contact regime as function of the magnetic field applied normal to the surface plane. The results are depicted in Figure 4.7. Overall, we notice that $Co(thd)_2$ shows a Kondo resonance at zero bias in both tunneling and contact conditions and one

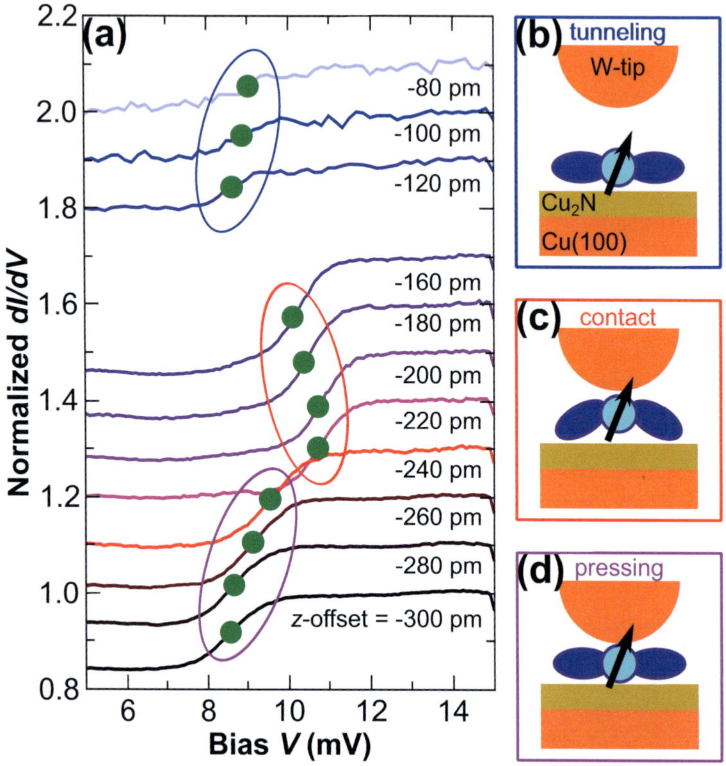

Figure 4.6: (a) Normalized dI/dV spectra on Co(thd)$_2$ measured at indicated z-offsets from a common initial set point (20 mV, 20 pA). The spectra are vertically offset by 0.1 at each step. Green dots illustrate the positions of the inelastic excitation in each spectrum. The blue, red and pink ellipses illustrate the tunneling, contact and pressing conditions, respectively. (b-d) Cartoons of tip on M(thd)$_2$ molecule in tunneling, contact and pressing conditions, respectively. Spectra were measured at 900 mK with a modulation of 0.2 mV RMS at 487 Hz.

magnetic excitation, while Ni(thd)$_2$ shows no Kondo effect but two excitations. Finally, Cu(thd)$_2$ again displays a Kondo resonance. Further, a weak inelastic step that is independent of magnetic field is present, which is due to a non-magnetic excitation. Since these ions share the same oxidation state of +2, the expected spins S for Co^{2+}, Ni^{2+} and Cu^{2+} are $3/2$ (4 states), 1 (3 states) and $1/2$ (2 states), respectively. We use these spins of the ions to explain in detail the observed behaviour as a function of magnetic field.

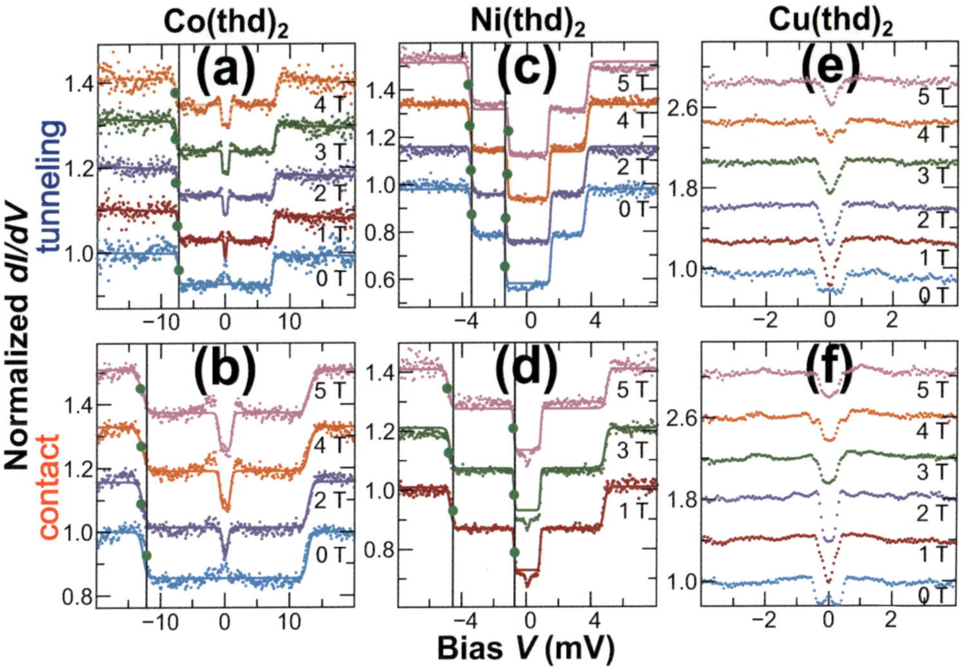

Figure 4.7: Field dependent dI/dV spectra. (a-f) present dI/dV spectra under various out of plane magnetic fields measured on Co(thd)$_2$, Ni(thd)$_2$ and Cu(thd)$_2$ in both tunneling (set point: 20 mV, 200 pA) and contact (set point: 20 mV, 2.5 nA) conditions, respectively. Solid lines on the pectra of Co(thd)$_2$ and Ni(thd)$_2$ represent the fits with a spin Hamiltonian as illustrated in the text. Neighboring spectra are vertically offset by 0.10, 0.16, 0.16, 0.20, 0.30 and 0.40 in panel (a-f) respectively. Green dots indicate the positions of the inelastic excitation on each spectrum (mid of the step). Vertical lines through excitation steps of lowest spectra are guides to the eye for the shifts of excitation energies. Spectra were measured at 30 mK with a modulation of 0.1 mV RMS at 2.21 kHz

4.1.6 Models of spin states

To extract the ligand field parameters, we fitted the spectra to a spin Hamiltonian, which can be represented by a sum of Stevens' operators respecting the C_{1v} symmetry (see Section 2.3) and the Zeeman energy of the external magnetic field. The Hamiltonian is:

$$\mathscr{H}_{\text{eff}} = -g\mu_B B_z S_z + 3B_2^0 S_z^2 + \frac{B_2^2}{2}(S_+^2 + S_-^2) + \frac{B_2^1}{4}[S_z(S_+ + S_-) + (S_+ + S_-)S_z], \qquad (4.1)$$

where g is gyromagnetic factor, μ_B is the Bohr magneton, B_z is the magnetic field, S_z, S_+ and S_- are the spin operators (z-component, raising and lowering operator), B_2^0, B_2^2 and B_2^1 are the axial and transversal magnetic anisotropy and the anisotropy term due to the low symmetry, respectively. Here, the z-axis is normal to the plane and the x-axis is in the mirror plane of the molecule. The tunneling spectra were calculated from the transitions between the eigenstates of $\mathscr{H}_{\text{eff}}$ by spin flip excitation, as discussed in detail in Ref [77]. This model describes both the position and height of the steps in the dI/dV spectra as a function of the coefficients of the corresponding $\mathscr{H}_{\text{eff}}$ and the g-factor, but does not capture Kondo physics. Thermal broadening and thermal population of the initial states were considered (see Section 2.6.2). All spectra of the individual species were fitted for tunneling and contact conditions simultaneously as the function of magnetic field (global fit). The fit to the spectra of Co(thd)$_2$, Ni(thd)$_2$ displayed in Figure 4.7 describes the spectra well. It naturally fails to describe the Kondo resonance and the overshoots in dI/dV near the steps. The coefficients of the spin Hamiltonians and the g-factor determined by fitting the field dependent spectra are listed in Table 4.2. Thanks to these known coefficients of the Hamiltonians, the corresponding eigenstates and energies of both tunneling and contact case can be solved, which are shown in Table 4.3. Consequently, in turn, the energy sub-levels of spin states can be established precisely to interpret the features observed in dI/dV spectra (see Figure 4.8). The connection between the model of spin states and the characteristic of the field dependent dI/dV spectra in both tunneling and contact conditions will be discussed below in detail for Co(thd)$_2$, Ni(thd)$_2$, respectively.

Table 4.2: The coefficients of the spin Hamiltonians and the g-factor determined by fitting the field dependent spectra of Co^{2+} and Ni^{2+} in tunneling and contact conditions. Energies are given in meV.

	Co(thd)$_2$		Ni(thd)$_2$	
	tunneling	contact	tunneling	contact
g	1.50 ± 0.05	2.06 ± 0.07	2.09 ± 0.05	2.2 ± 0.2
B_2^0	-0.71 ± 0.05	0.3 ± 0.2	0.818 ± 0.001	0.64 ± 0.03
B_2^2	1.74 ± 0.06	1.5 ± 0.3	1.058 ± 0.004	2.26 ± 0.04
B_2^1	-0.003 ± 18	6.6 ± 0.2	-0.04 ± 0.1	0.73 ± 0.07

Table 4.3: Eigenstates and eigenenergies to spin Hamiltonians quantitated by fitting the spectra.

		eigenstate	eigenenergy (mV)
Co(thd)$_2$	tunneling	$0.4601\,\|1/2\rangle + 0.0004\,\|-1/2\rangle - 0.8879\,\|-3/2\rangle$	-6.36
		$0.8879\,\|3/2\rangle + 0.0004\,\|1/2\rangle - 0.4601\,\|-1/2\rangle$	-6.36
		$0.8879\,\|1/2\rangle + 0.0008\,\|-1/2\rangle + 0.4601\,\|-3/2\rangle$	1.03
		$-0.4601\,\|3/2\rangle + 0.0008\,\|1/2\rangle - 0.8879\,\|-1/2\rangle$	1.03
	contact	$0.3147\,\|1/2\rangle - 0.6847\,\|-1/2\rangle - 0.6573\,\|-3/2\rangle$	-5.29
		$-0.6573\,\|3/2\rangle + 0.6847\,\|1/2\rangle + 0.3147\,\|-1/2\rangle$	-5.29
		$0.2745\,\|1/2\rangle - 0.5973\,\|-1/2\rangle + 0.7536\,\|-3/2\rangle$	7.46
		$0.7536\,\|3/2\rangle + 0.5973\,\|1/2\rangle + 0.2745\,\|-1/2\rangle$	7.46
Ni(thd)$_2$	tunneling	$-0.0106\,\|1\rangle - 0.9998\,\|0\rangle + 0.0106\,\|-1\rangle$	-0.0003
		$0.7070\,\|1\rangle - 0.0150\,\|0\rangle - 0.7070\,\|-1\rangle$	1.40
		$0.7071\,\|1\rangle + 0.7071\,\|-1\rangle$	3.51
	contact	$-0.5956\,\|1\rangle + 0.5389\,\|0\rangle + 0.5956\,\|-1\rangle$	-0.57
		$0.3811\,\|1\rangle + 0.8424\,\|0\rangle - 0.3811\,\|-1\rangle$	0.23
		$0.7071\,\|1\rangle + 0.7071\,\|-1\rangle$	4.19

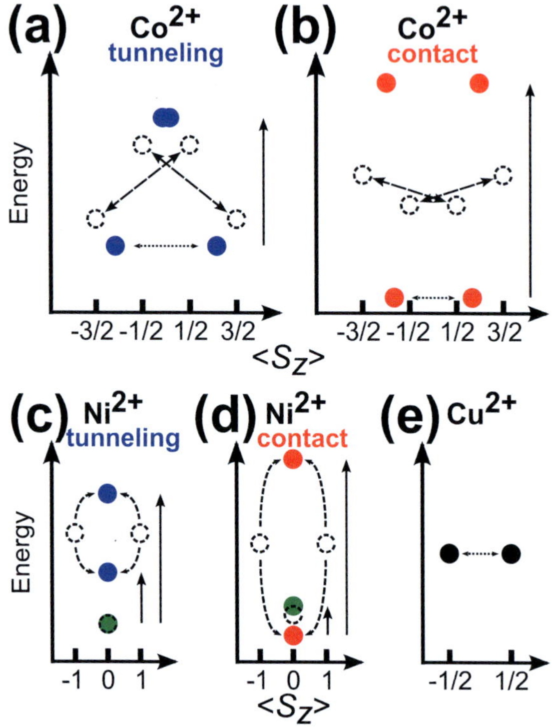

Figure 4.8: Sketch of ligand field models of spin states. Energy and $< S_z >$ of the eigenstate of Co^{2+} (a,b) and Ni^{2+} (c,d) in tunneling and contact conditions, respectively, and Cu^{2+} (e). Full dots represent eigenstates. Vertical solid arrows indicate spin excitations. Horizontal dotted arrows indicate Kondo resonances. Dashed circles represent the eigenstates of the B_2^0-term of the Hamiltonians. Dashed arrows in (a,b) show B_2^2-term mixing second neighbouring eigenstates of B_2^0-term. Green dots in (c,d) represent the state mainly composed from the $S_z = 0$ wave function. Bended dashed arrows in (c,d) illustrate that B_2^2 and B_2^1-terms split the doublet to two singlets with $< S_z >= 0$.

For Co^{2+}

For Co^{2+} with a half integer spin, the four magnetic states will be split by the ZFS into two Kramers doublets (see Figure 4.8(a,b)). We expect an inelastic excitation from the ground state doublet to the excited state doublet at a finite energy and a Kondo effect from elastic scattering of conduction electrons with the two states of the ground state doublet. These two features are clearly visible in the spectra (see Figure 4.7(a,b)). By the application of a magnetic field, the ground state doublet responsible for the Kondo effect splits into two states and the Kondo peak disappears. Instead, an inelastic excitation emerges due to spin flip scattering within the Zeeman split doublet. Similarly, the excitation from the lower doublet to the higher doublet is modified by the magnetic field. In both tunneling and contact conditions, the inelastic excitation shifts to higher energies with increasing field (see green dots and vertical black line).

Regarding the parameters extracted from the fitting, for the tunneling case, B_2^1 is compatible with zero, but the positions of the steps do only weakly depend on B_2^1 causing a large uncertainty for B_2^1. Also note that in this case, g is somewhat lower than 2 indicating some orbital contribution to the magnetic moment. B_2^0 is negative indicating an easy axis normal to the plane. When going from tunneling to contact condition, B_2^0, B_2^2, and B_2^1 change significantly. B_2^1 deviates significantly from zero and artificially fixing B_2^1 to zero in this case does lead to unphysical fits. Interestingly the sign of B_2^0 reverses upon contact formation, i.e. the anisotropy is now that of an easy plane system.

As depicted in Figure 4.8(a), the four initial magnetic states of the Co ion with $S_z = -3/2$, $-1/2, 1/2$ and $3/2$ are mixed by the Stevens' operators. The B_2^2-term mixes states with $\Delta S_z = 2$, i.e. S_z= +3/2 with -1/2 and +1/2 with -3/2, leading to a reduction of the expectation values of S_z. Further, the B_2^1-term mixes neighbouring states and the B_2^0-term is negative, i.e. overall a downward parabola is formed with a ground state doublet of high content of the $S_z = \pm 3/2$ states (see Table 4.3). As the spin is half integer, all eigenstates are Kramers doublets, i.e. two doublets result. Upon contact formation, the sign of B_2^0 reverses and by this, the energetic order of the two doublets reverses as well. The ground state doublet in contact has a higher content of $S_z = \pm 1/2$ (see Figure 4.8(b)). The Kondo effect in both cases is caused by transitions between the two states of the ground state doublet from elastic spin flip scattering by a single substrate or tip electron. This is allowed since the ground state doublet contains components with $S_z = \pm 1/2$ in both cases (see Table 4.3).

For Ni^{2+}

For Ni^{2+} with an integer spin, the three magnetic states will be split by the ZFS into three singlets (see Figure 4.8(c,d). Note that the magnetic states with $S_z = +1$ and -1 are coupled by the B_2^2-term. As the doublet is not of a Kramers kind, this coupling will lead to a splitting of the doublet and formation of two singlets with vanishing expectation value of S_z (for the composition of the states see Table 4.3). Excitations from the ground state singlet are possible to both excited states and we therefore observe two inelastic excitations. As there is only one ground state, no Kondo effect is observed. Note that in tunneling conditions, the lower energy excitation shifts to lower energies and the higher excitation to higher energies with rising field (see green dots in Figure 4.7(c)). In contact, however, both excitation energies rise with field (cf. Figure 4.7(d)). We further observe a minor dip at zero bias, which does not depend on magnetic field. Thus, this particular excitation cannot be of magnetic origin and most likely it is caused by a vibronic excitation. Further, we note that the height of the excitations steps in the spectra is in general larger than in case of Co. Similar to the case of Co, contacting the molecule modifies the ZFS. The fits to the spectra reveal again a g-factor close to 2 and a drastic increase of B_2^1 upon contact formation. As depicted in Figure 4.8(c,d), the order of the states changes upon contact formation, similar to the case of Co^{2+}. While in the tunneling regime, the ground state is mainly composed from the $S_z = 0$ wave function, it is moved upward between the two other states when the molecule is contacted. In this situation, the application of a magnetic field will lead to an increase of both excitation energies, in agreement with the experimental observation.

For Cu^{2+}

Finally, we discuss the case of Cu^{2+}, which should display a spin of $1/2$. In this case, one obtains a Kramers doublet, all Stevens' parameters vanish, and the Hamiltonian only contains the Zeeman energy (see Figure 4.8(e)). In agreement to that, we observe a Kondo peak at zero bias, which splits into two steps with the magnetic field (see Figure 4.7(e,f)). In addition to the magnetic excitation, there is a vibronic excitation, which does not move with magnetic field.

4.1.7 Partial summary and outlook

Thus, our results indicate that the ZFS can be drastically modified by contacting or breaking a contact and hence suddenly deforming the molecule. Care has to be taken when extrapolating the ZFS obtained from crystals of molecules to break junction experiments. Even opening and closing the junction can fully reverse the sign of the leading anisotropy term and exchange the energetic order of the spin states.

These observations indicate to take great care of the microscopic deformation in magnetic molecules especially in break junction experiments but also offers ways to manipulate the quantum states of molecules. For example, an excitation of the break junction with acoustic waves in the GHz range would allow to use the coupling between mechanical and spin degrees of freedom to manipulate the spin state of the molecule. Note that the selection rules for spin excitations driven by the modification of the ZFS and driven by microwave absorption [78] are different opening new ways to prepare quantum states.

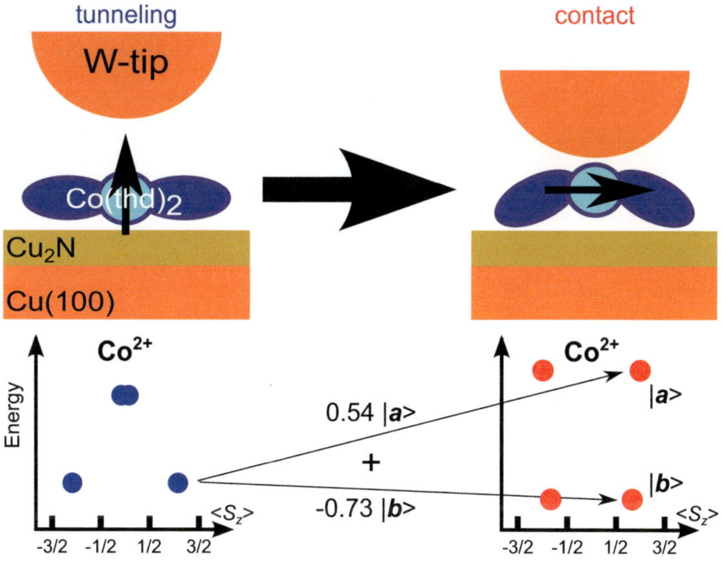

Figure 4.9: A sketch of preparing a superposition state. The upper cartoon demonstrates the abrupt formation of the molecular contact. The arrows in the lower panel illustrate the corresponding transformation from one state of the ground doublet mainly consisting of the $S_z = 3/2$ state in the tunneling regime to a superposition state mainly consisting of two new eigenstates $|a\rangle$ and $|b\rangle$ of the contact condition with coefficients of 0.54 and -0.73, receptively.

Further, the abrupt nature of contact formation can potentially be used to prepare superposition states [63, 79, 64]. If we assume that the molecular spin is in one of the ground state when contacted (i.e. the contact is kept for times longer than the lifetime of all excited states), the abrupt modification of the ZFS by contacting the molecule with the tip will move this state to a higher energy, i.e. changes the energetic order of the states. This sudden transformation will change the composition of the states according to time dependent perturbation theory, even in the absence of a tunneling current. If then the tip is retracted again, the prior order of the state is restored. During one process of modifying the ZFS, a transition between the states will thus be induced, which in principle can be used to prepare a superposition final state. For instance, if the Co^{2+} ion in the tunneling regime stays in one state of the ground doublet mainly consisting of the $S_z = 3/2$ state, after the abrupt formation of contact, the same spin state will transform to a superposition state mainly consisting of two new eigenstates in the contact regime (see Figure 4.9). For this to function, the changes in ZFS need to be non-adiabatic. To effectively prepare these superposition states, a slight modification in the energy of the ground state is not sufficient, the energetic order of the states needs to be altered by contact closing (or opening). The involved time scale of snapping closing of molecular contact is roughly given by the inverse phonon frequencies [80], i.e. is of the order of ps. The involved time scale in molecular spins, i.e. the coherence time, is often in the range of 100 ns to 10 µs [81]. Under these assumptions, the final quantum state after closing the junction is a superposition state that can be calculated by projecting the initial state with open contact onto the final states of the closed contact, which is, for example, $0.54\,|a\rangle - 0.73\,|b\rangle$ in the case of Figure 4.9. Unfortunately, the coherence times of molecular spins in this experiment are not long enough for our STM to detect the superpositions created by opening or closing the contact. However, one can imagine this mechanism to work in systems with long coherence times. Alternatively, one may use surface acoustic waves in the GHz frequency range to induce deformations and perturbations of the ZFS Hamiltonian to coherently induce transitions in the multiplet.

The work of this section has been accepted by ACS Nano for publication.

4.2 MnPc molecules

Phthalocyanine (Pc) with the formula $(C_8H_4N_2)_4H_2$ was first reported in 1907 [82]. It is a simple metal-free organic compound. However, its metal complexes (MPc) have broad applications, which include not only daily consumables such as dyes, pigments, and catalysts [83], but also high-tech electronics including organic field-effect transistors, diodes, photovoltaics and gas sensors [84].

Since different metal elements can be employed as a central ion, the series of MPc molecules exhibit rich properties. Additionally, MPc molecules are impressively stable, i.e. rarely decompose below 900 °C in vacuum. Therefore, they are frequent research objects in the topic of molecules on surfaces, especially when STM is used. It has been suggested by first-principles calculations that each manganese phthalocyanine (MnPc) molecule carries a magnetic moment of $\approx 3.0\ \mu_\mathrm{B}$ [85]. Hence, we were motivated to investigate MnPc molecules with our STM at 5 K. In this project, Cu(100) and Co/Cu(100) were used as substrates to deposit MnPc molecules. The details about sample preparation are introduced in Section 3.2.

The structural formula of MnPc is sketched in Figure 4.10. The Mn ion in the centre is surrounded by four isoindole-like units. The whole square planar molecule presents a $D_{4\mathrm{h}}$ symmetry.

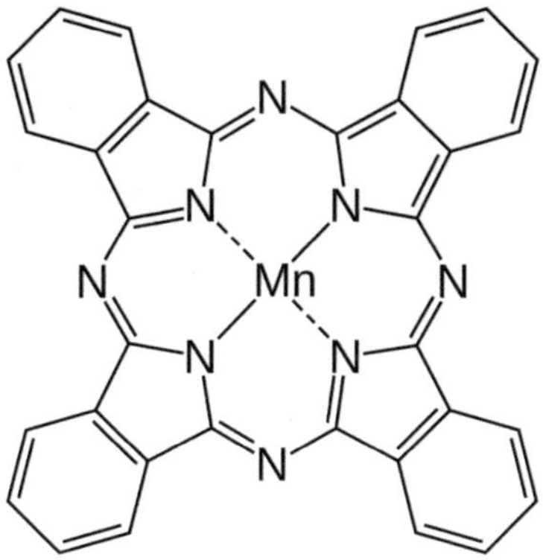

Figure 4.10: Structural formula of manganese phthalocyanine.

4.2.1 MnPc on Cu(100)

Topography

Figure 4.11 shows the topography of a sub-monolayer of MnPc molecules deposited on Cu(100). On the flat surface, besides irregular contaminants, individual protrusions with a bright centre and cross-like outline (illustrated by two white cross frames) can be easily recognized, which refer to single MnPc molecules lying flat on the surface. This characteristic shape, which is consistent with the structure of the molecule, has been observed for many MPc molecules [86, 87].

The Kondo resonance

To confirm the magnetic properties, we measured dI/dV spectra on MnPc molecules. As shown in Figure 4.12, in contrast to the flat background taken on bare Cu(100), the spectrum taken on the centre of a MnPc displays a pronounced peak near zero bias voltage. This peak spans over more than 100 mV. Actually, this resonance peak is reproducible on every MnPc molecule. Although, we could not examine the temperature and external magnetic field dependences of this feature, we deduce that it resulted from a Kondo resonance, since similar behaviours were evidenced on CoPc molecules deposited on Au(111) as well as MnPc or CoPc deposited on Pb/Si(111) by STS [88, 89, 51]. Similar to Kondo resonance observed on $Co(thd)_2$ in last section, the peak here was fitted to the same Fano function

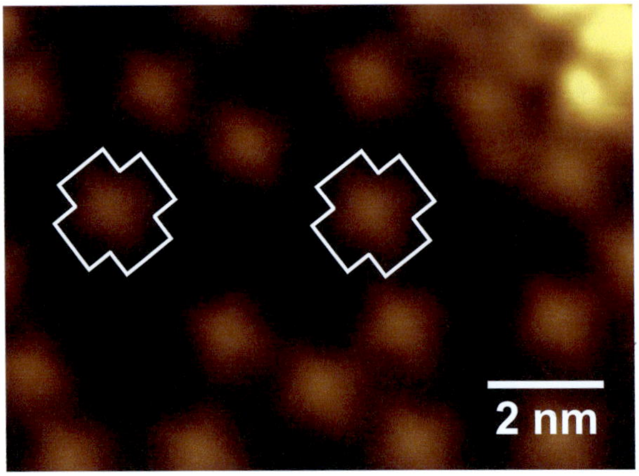

Figure 4.11: STM topography of sub-monolayer of MnPc molecules on Cu(100). The cross-like shape of molecules is highlighted by two white frames. Set point: -0.6 V, 50 pA.

(see Section 2.7.3) as shown by the green line in Figure 4.12. The fitted parameters of the Fano function are displayed in Table 4.4). Note that the extracted Kondo temperature T_K is as high as 207 ± 6 K. This impressive value agrees with the reported T_K of 208 K for CoPc on Au(111) [88], which was attributed to a small on-site Coulomb repulsion U and a large half-width of the hybridized d-level Δ (in consistency with the expression of T_K described in Equation 2.27). We deduce that the high T_K for MnPc on Cu(100) can be interpreted by same mechanism.

Table 4.4: Determined coefficients of fitting Kondo resonances near zero bias on dI/dV spectra of MnPc on Cu(100) in Figure 4.12 with a Fano function: $y(x) = R_0(q + \frac{x - E_K}{k_B T_K})^2/(1 + (\frac{x - E_K}{k_B T_K})^2) + kx + y_0$, where T_K, q and E_K are the Kondo temperature, Fano parameter and position of Kondo resonance, respectively, $kx + y_0$ is a linear background.

T_K (K)	207 ± 6
q	0.17 ± 0.02
E_K (mV)	-0.0009 ± 0.0006
R_0	-0.211 ± 0.003
k	0.51 ± 0.05
y_0	0.95 ± 0.003

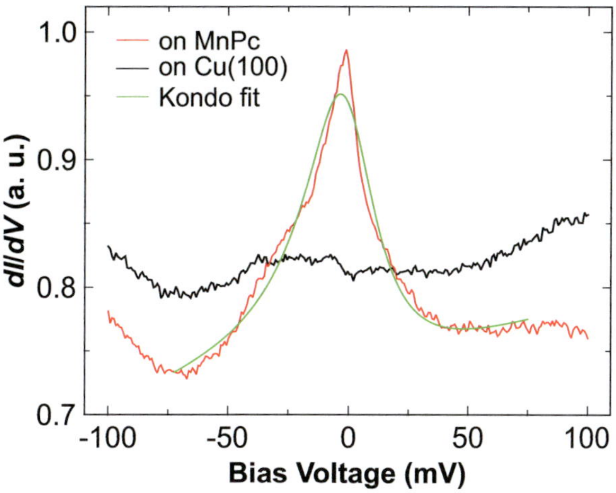

Figure 4.12: Comparison of dI/dV spectra taken on single MnPc molecule (red) and on bare Cu(100) substrate (black). Lock-in modulation amplitude: 2 mV, frequency: 2.81 kHz. Setpoint: 0.1 V, 500 pA. The green line refers to the fit of a Fano function.

4.2.2 MnPc on Co/Cu(100)

It is known that planar MPc molecules prefer to stack in columns in a film thicker than one monolayer (ML). The staking geometry plays a key role for the physical properties of the film, e.g. the magnetic correlations between MnPc molecules from different layers [90, 91, 92]. Our cooperators observed a robust antiferromagnetic order of staking layers in MnPc films with X-ray magnetic circular dichroism (XMCD) [93]. Interestingly, the first-ML MnPc was ferromagnetically coupled to the underlying Co layer thence inducing an exchange bias on Co at low temperatures. Separately, the *ab initio* density functional theory calculations demonstrated that the staking geometry strongly influences the strength of the magnetic couplings [93]. In comparison of three different relative adsorption geometries between the first- and the second-ML of MnPc (see Figure 4.13), the calculated total energies for both ferromagnetic and antiferromagnetic coupling, as well as other parameters are listed in Table 4.5. Note that the total energy varies slightly for different configurations, in spite that an antiferromagnetic coupling is favoured for all of three configurations.

To experimentally examine the staking geometry of the MnPc films, X-ray natural linear dichroism (XNLD) is a powerful technique, since it can detect the orbital anisotropy when all of the molecules share one orientation. For instance, by searching the direction of the unoccupied π^* orbitals, a geometry of MnPc molecules lying either flat or vertically on the surface can be uncovered. However, in the former case, the in-plane orientations of molecules are difficulty for XNLD to detect, whereas STM is an unparalleled technique to reveal the adsorption geometries of individual molecules on surfaces. In the following, the topography and dI/dV spectra of MnPc molecules deposited on ≈ 7 ML of Co grown on Cu(100) will be presented. The growth of Co layer is introduced in Section 3.2.

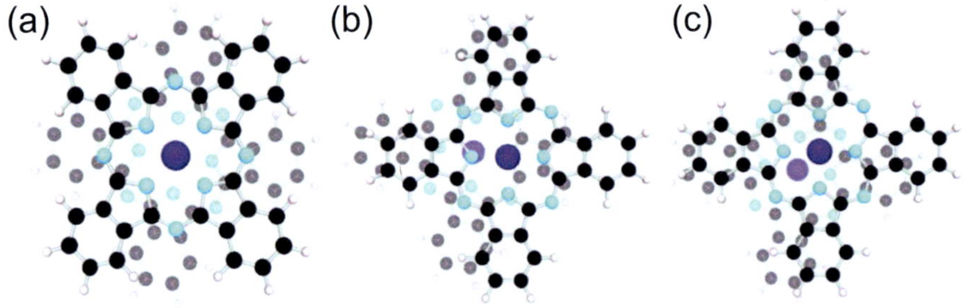

Figure 4.13: Three different adsorption geometries calculated for the second-ML MnPc. (a) the second-ML molecule sits on top of the first monolayer, rotated by 45°. (b) the second monolyer is laterally shifted such that the second-ML Mn ion (purple dot) is positioned on top of an underlying N ion (cyan dot). (c) the second-ML molecule is laterally shifted compared to the first-ML molecule, such that the Mn ion sits in a hollow position. This figure is reprinted with permission from Ref. [93].

Table 4.5: For each of the three different adsorption geometries demonstrated in Figure 4.13, *ab initio* calculations yield: the total energy of ferromagnetic (antiferromagnetic) configuration $E_{\mathrm{FM(AF)}}$, the spin magnetic moment $m_s^{\mathrm{Mn}_i}$ ($i = 1,2$ for the ion in the first- or second-ML, respectively), the distance $d_{\mathrm{Mn}_1-\mathrm{Mn}_2}$ between the Mn sites of first- and second-ML, and the distance Δz perpendicular to the surface between the two Mn atoms. The figure is taken from Ref. [93].

	Config. (a)	Config. (b)	Config. (c)
E_{FM} (eV)	−2081.873	−2081.879	−2081.858
E_{AM} (eV)	−2081.906	−2081.884	−2081.865
$E_{\mathrm{FM}} - E_{\mathrm{AM}}$ (mV)	33	5	7
$E_{\mathrm{FM}} - E_{\mathrm{AM}}$ (K)	384	59	79
$m_s^{\mathrm{Mn}_1}$ (μ_{B})	2.51	2.54	2.53
$m_s^{\mathrm{Mn}_2}$ (μ_{B})	−3.04	−3.045	−3.05
$d_{\mathrm{Mn}_1-\mathrm{Mn}_2}$ (Å)	3.87	4.36	4.22
Δz (Å)	3.87	3.87	3.87

The adsorption geometry

As shown in Figure 4.14, on large-scale flat Co terraces, the MnPc molecules deposited on the surface with a sub-monolayer coverage lie flat with a four-fold cross-like shape (see the magnification of one molecule), which is similar to the case on Cu(100) (see Section 4.2.1). The [011] crystallographic orientation (white arrow) can be deduce from the predominant edges of Co islands grown on *fcc* Cu(100). With respect to the direction, the four-fold symmetric MnPc molecules mainly presents two in-plane orientations, which are indicated by red and green crosses, respectively.

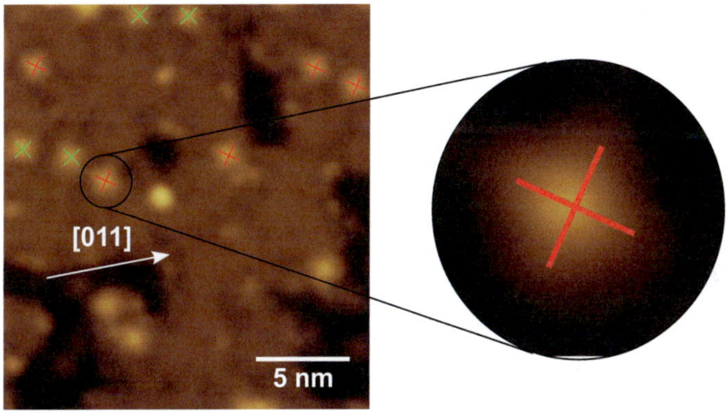

Figure 4.14: Topography of MnPc molecules on Co/Cu(100) with a low coverage. One molecule is zoomed with an enhanced contrast. The [011] crystallographic direction of the substrate is indicated by the white arrow. Molecules with two different in-plane orientations are marked by red and green crosses, respectively. Set point: 1 V, 100 pA.

When the coverage of MnPc molecules was increased to $\approx$ 1.5 ML, we observed an almost completed first monolayer, a half completed second monolayer consisting of isolated molecular islands on which third monolayer already started to stack (see Figure 4.15(a)). The three layers can also be recognized in a line profile presented in Figure 4.15(b). Remarkably, in contrast to the first-ML molecules presenting a simple cross-like shape with four simple lobes, the second- and third-ML molecules display a more complicated structure referring to the molecular orbitals (see the magnification of one molecule), which is similar to the behaviours of FePc and YPc_2 on Au(111) [94, 95]. The appearance of the details of the molecular orbitals is ascribed to the decoupling between the second or third monolayer from the metallic substrate by the first monolayer. However, these molecules still lie flat and an overall four-fold symmetry is preserved.

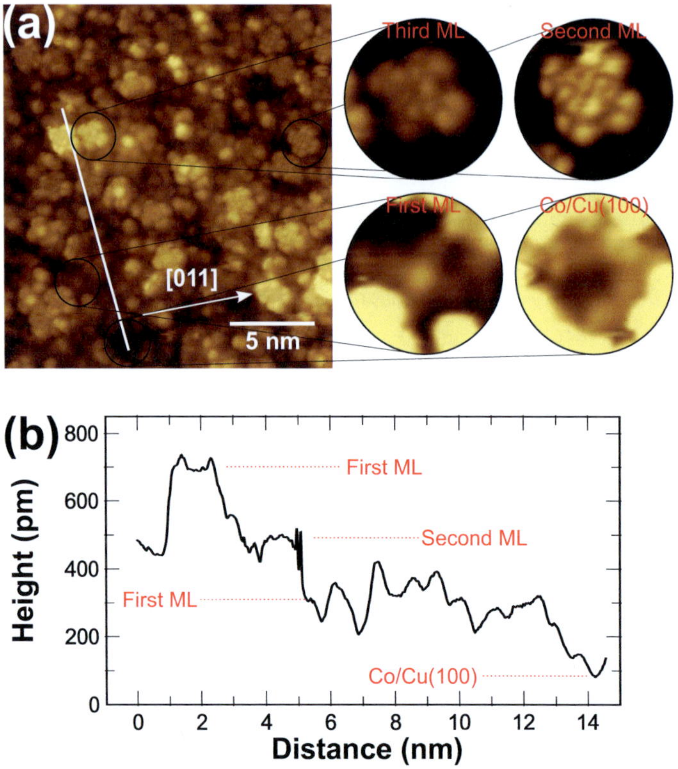

Figure 4.15: (a) STM image of MnPc molecules on CO/Cu(100) with a $\approx$ 1.5 ML coverage. In the centres of four zoomed areas with normalized z scale, the topographies of three molecules from third, second and first monolyer as well as bare Co/Cu(100) are presented. A white arrow indicates the [011] direction. Set point: 1 V, 50 pA. (b) a line profine along the white line in (a). Four different levels in height are lablled.

Note that molecules from all of the monolayer show various in-plane orientations. We statistically analysed the orientations of molecules from different monolayers separately in a larger area (see Figure 4.16(a)). The orientation for each molecule relative to [011] direction of Co/Cu(100) was extracted in the following processes: first, each molecule was marked by a cross whose two long axes were parallel to the lobes of the molecules. One example is shown by the red cross in the magnification of Figure 4.14. Subsequently, the orientation angle is defined by the smaller one between two positive angles from the [011] direction to both long axes of the remarking cross. Note that the two axes of the cross are equivalent due to the four-fold symmetry of the molecules. Molecules in third, second and first monolayer are marked by black, blue and white crosses, respectively. The distributions of the orientation angles of molecules in different monolayers are shown in three histograms (see Figure 4.16(b)). According to the histograms, the preferred orientation angles for each monolayer are concentrated in two similar small ranges within 10°, which are near 28° and 52°, 23° and 54 , 22° and 55 for third-ML, second-ML and firs-ML respectively. Note that due to the four-fold symmetry for both MnPc molecules and the substrate, the coexistence of two preferred orientation angles for every monolayer is reasonable, which should perpendicular to each other. With respect to these distributions, we refer that planar MnPc molecules in different monolayer have identical in-plane orientations, namely, there is no relative rotation between two staking layers. Unfortunately, since the molecules are not well organized, the relative lateral shifts of Mn ions in different layers cannot be checked by our STM topographies.

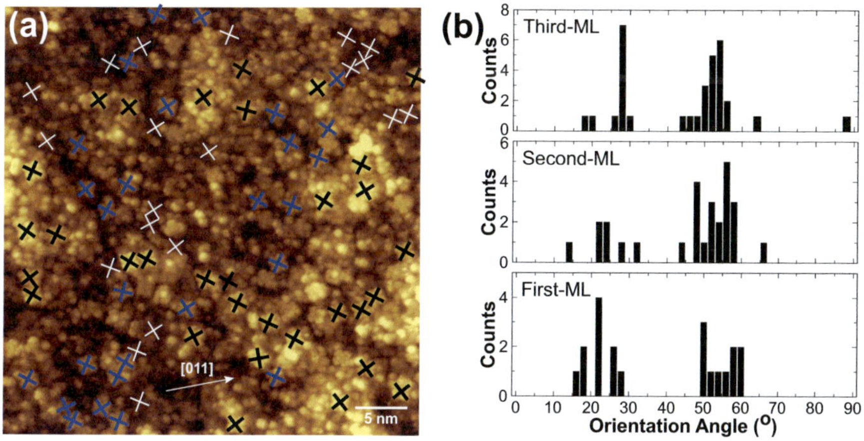

Figure 4.16: (a) topography of a larger area. The orientations of molecules are demonstrated by the black, blue and white cross for the first, second and third monolayer, respectively. A white arrow indicates the [011] direction. Set point: 1 V, 50 pA. (b) histograms of the distributions of the orientation angles of molecules in different monolayers.

The adsorption geometries of MnPc molecules on Co/Cu(100) has been published as one part of Ref. [93].

The inelastic excitation

dI/dV spectra were also measured in the centres of MnPc molecules in the second monolayer on Co/Cu(100) (see Figure 4.17). As expected, the resonance peak that was observed in the case of MnPc on Cu(100) disappeared here, which is ascribed to annihilation of the Kondo state due to the presence of the ferromagnetic Co layer. However, a dip in the range from −40 mV to 40 mV is distinct in contrast to the background taken on bare Co. We attribute this phenomenon to the spin-flip induce inelastic magnetic excitation, which is also discussed in the case of M(thd)$_2$ molecules in Section 4.1. Similar inelastic excitations were observed for CoPc molecules on Pb/Si(111) [51].

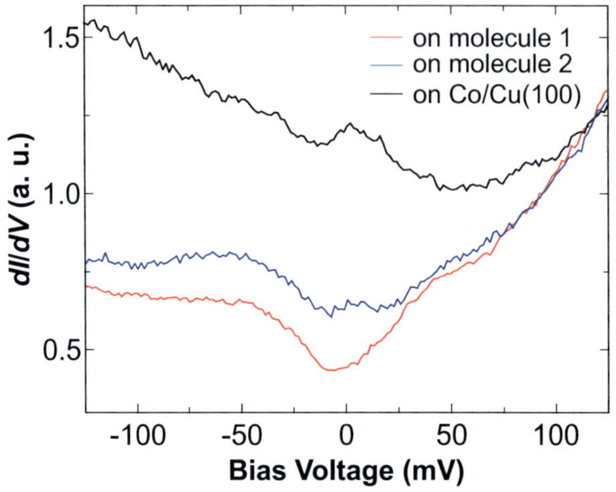

Figure 4.17: Comparison of dI/dV spectra taken on the centres of two MnPc molecules (red and blue) and on bare Co/Cu(100) substrate (black). Lock-in modulation amplitude: 2 mV, frequency: 2.81 kHz. Setpoint: 0.1 V, 100 pA.

Partial summary

In this section, the topographies and dI/dV spectra of MnPc molecules deposited on Cu(100) and Co/Cu(100) were studied. The adsorption geometries revealed that the planar molecules predominantly lie flat and stack layer-by-layer (at least for first three monolayers). Thanks to the decoupling effect of first monolayer, the structure of the molecular orbitals for second and third monolayer was depicted in STM images. Regarding the in-plane orientation with respect to the [011] direction of the substrate, two perpendicular directions are favoured by all of the molecules from three monolayers. Separately, in the dI/dV spectra, a pronounced Kondo resonance with T_K as high as 207 K was observed for MnPc molecules on Cu(100). A spin-flip inelastic excitation was revealed for MnPc on Co/Cu(100). The results of this project help to understand the the interfaces between organic molecules and the underneath metallic substrates, which is crucial to design organic spintronic devices, especially when the substrates are ferromagnetic which are usually referred to spinterface.

5 Magnetic molecules with 4*f* ions: $Tb(thd)_3$ and $[Tb(tmhd)_3]_2bpm$

Compared to the 3*d* series discussed in previous chapter, the complexes with 4*f* (lanthanide) ions present at least two attractive advantages as potential single magnetic molecules (SMMs): larger magnetic momenta and potentially longer magnetization relaxation times. The latter results from a larger magnetic anisotropy due to a stronger spin-orbit coupling (SOC) together with a crystal field (i.e. zero-field splitting (ZFS) as discussed in Section 2.2). In other words, the energy difference between the ground state and the first excitation state is larger. The stronger SOC can be explained by unquenched orbital angular momenta of 4*f* ions, in comparison to the quenched ones of 3*d* ions. Actually, SMMs with 4*f* ions came out earlier than that with 3*d* ion. They have attracted much interest due to the huge expected energy barrier for magnetization reversal [96], which is an an advantage from an applications point of view for magnetic storage devices based on single molecules. It has been reported that lanthanide-based SMMs present a slow relaxation with energy barriers of the the order of 1000 K as well as a retained hysteresis at liquid hydrogen temperature [97, 98], which is much higher than the barriers observed for SMMs based on 3*d* ions [99]. To date, almost all of the reported magnetic molecules with single 4*f* ion are based on terbium and dysprosium [100]. These two elements are common options to maximize the effect of ZFS on the magnetic behaviour of a lanthanide-based SMM. Particularly, complexes containing single terbium ions present highest magnetization relaxation barriers.

Scanning tunneling microscopy (STM) is one of the few methods that can reveal the magnetic properties on the level of single molecules, which has been widely used to investigate the magnetic properties of single ions and single molecules of transition metals [101, 102, 53, 103]. However, most of the attention is attracted by complexes based on the first-row transition metals. There are very few studies on lanthanide-based SMMs with low-temperature STM during the latest decade [104, 105], despite the variety of reports of their peculiar magnetic properties evidenced by alternating current magnetic susceptibility. Moreover, most of the reported STM studies of complexes with single lanthanide ion focus on double-decker bis(phthalocyaninato)lanthanide ($LnPc_2$), which contains a lanthanide ion sandwiched by two Pc rings. Kondo resonances have been observed on $LnPc_2$ [104, 106], while spin-flip excitation on any lanthanide-based SMM has not been evidenced. The field of investigating lanthanide-based SMM with STM is still young, but it is definitely worth the effort.

One plausible reason for this small quantity of studies is the difficulty of transferring the molecules onto substrates in an ordered and clean way (cf. $LnPc_2$ molecules are very stable). Because of the huge molecular mass of lanthanide-based SMMs, many of them cannot be sublimed without decomposition, even in vacuum. Some studies have suggested methods other than vacuum deposition to solve the problem, but they are limited by the resulting level of contaminations of the surface [107]. Although $LnPc_2$ molecules show novel magnetic properties and are easy to deposit onto surfaces, more potential magnetic molecules with single $4f$ ion are desired. Finding other classes of rare-earth SMMs that can be transferred would be a significant step forward towards the application of magnetic molecular devices in which well-defined homogeneous molecular structures are required. To achieve that we tested lanthanide centred molecules with other ligands rather than Pc rings.

In this chapter, as an example of molecules with single lanthanide ions, the investigation of tris(2,2,6,6-tetramethyl-3,5-heptanedionato)terbium(III) ($Tb(thd)_3$) with STM will be presented. The $Tb(thd)_3$ molecules were deposited onto various metal surfaces, thence different reconstructions of the molecules were formed due to different molecule-substrate interactions. Scanning tunneling spectroscopy revealed the distributions of molecular orbitals under particular bias voltages. Besides the mono-nuclear lanthanide complex $Tb(thd)_3$, another binuclear lanthanide complex $[Tb(tmhd)_3]_2bpm$ (tmhd=2,2,6,6-tetramethyl-3,5-heptanedionate, bpm=2,2'-bipyrimidine) was deposited onto a Au(111) substrate and thence studied by STM. $[Tb(tmhd)_3]_2bpm$ molecules were assembled to form a well-organized film with hexagonal patterns. Our results imply that these lanthanide complexes can be transferred onto the metal substrates without molecular decomposition or contaminations on the surface. These new rare-earth-based classes of molecules broadens the choice of molecular magnets to be studied with STM. Note that the main results of this chapter have been published in Ref. [108] and Ref [109].

5.1 Tb(thd)$_3$

The β-dikenonate lanthanide(III) series have been known as volatile lanthanide complexes since the 1970s and have been used as precursors in chemical vapor deposition in the thin-film industry [110]. However, to the best of our knowledge, there has been only one paper that reports on the characterization of isolated molecules of ruthenium (platinum group metal) β-dikenonate with low-temperature STM [111].

Tris(2,2,6,6-tetramethyl-3,5-heptanedionato)lanthanide(III) (Ln(thd)$_3$) has the simplest molecular structure among β-dikenonate lanthanide and is highly volatile and thermally stable [112, 113]. In this molecule, the lanthanide(III) ion is coordinated by three β-dikenonate ligands (see Figure 5.1) and the total charge of the molecule is zero. Recent DFT calculations for Ln(thd)$_3$ in gas phase show that the D_3 symmetry structure corresponds to the minimum of the potential energy [114]. Though remarkable magnetic properties of Ln(thd)$_3$ in bulk have not been reported, some of their derivatives exhibit SMM behavior [115]. Thus, it is interesting to investigate Ln(thd)$_3$ with low-temperature STM.

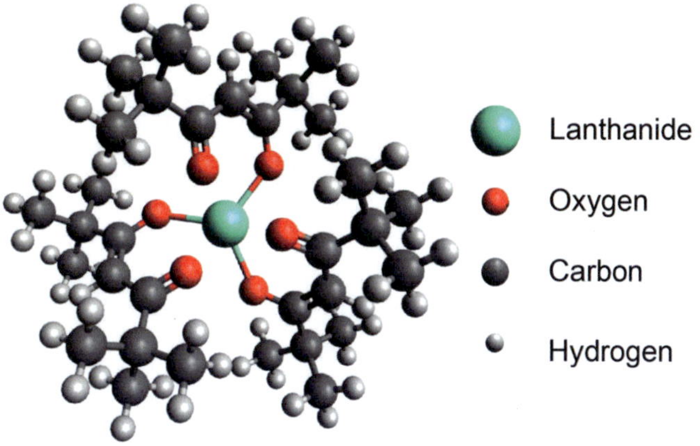

Figure 5.1: A ball-stick model of the tris(2,2,6,6-tetramethyl-3,5-heptanedionato) lanthanide(III) molecule.

In this project, in order to explore the transferability of lanthanide molecules onto metal surfaces without decomposition, Tb(thd)$_3$ (Tb = terbium) molecules with 99 % purity purchased from Alfa Aesar were deposited onto Au(111), Ag(111) and Cu(111) surfaces from a ceramic crucible at $\approx 62\,^\circ$C (more detail about sample preparation has been introduced in Section 3.2). The measurements were preformed in the JT-STM (see Section 3.1.1). During the measurement, the sample temperature was kept at ≈ 5 K. The dI/dV spectra were taken using a standard lock-in amplifier technique (see Section 3.3.2) with a 487 Hz modulation frequency and 20 mV modulation voltage. The dI/dV maps were recorded with the same lock-in parameters but with running feedback loop.

5.1.1 Configurations of Tb(thd)$_3$ molecules deposited on various substrates

STM topographies of Tb(thd)$_3$ deposited on Cu(111), Ag(111) and Au(111), respectively are presented in Figure 5.2(a-f). The white arrows indicate the $[1\bar{1}0]$ direction, which was determined from images with atomic resolution of each surface. We observed large molecular islands of Tb(thd)$_3$ growing homogeneously on Cu(111) and Ag(111) but very few isolated molecules on these substrates (see Figure 5.2(a,c)). Actually, the deposited molecules diffuse at room temperature on the substrate. Subsequently, nucleation of islands occurs due to an attractive molecule-molecule interaction. Eventually, a sub-monolayer film forms. The height of an island relative to the substrate in the topography sensitively depends on the sample bias, whose value is $\approx 340\,\mathrm{pm}$ under a bias of 1 V. In stark contrast, on Au(111), we observed isolated molecules and small clusters at the elbow sites of the Au(111) herringbone reconstruction [116], but no large islands (see Figure 5.2(e)). Zooming in, an isolated molecule in the center and two clusters containing three molecules (i.e. molecular trimers) are shown in Figure 5.2(f). The formation of the clusters at the elbow sites of the herringbone implies a higher adsorption energy at these sites comparing to that at the other sites. Hence, the molecules are trapped but cannot diffuse further on the surface to form larger islands at room temperature. This preferential nucleation at the elbow sites has been reported for other molecules [117, 118].

5.1.2 Superstructures of the molecular films

In the extended islands on Cu(111) and Ag(111), a quasi-triangular lattice composed of molecules was observed (see Figure 5.2(b,d)). On Cu(111), we observed a particular row-like structure: two-molecule-wide ribbons separated by dark stripes were formed along a direction which is 11° rotated from $[1\bar{1}0]$ (see Figure 5.2(b)). Looking at Figure 5.2(a) closely, we find two species of film reconstructions, labelled by R and L, which are $-11°$ and 11° rotated from the $[1\bar{1}0]$ direction, respectively. These two species are linked by a mirror operation along $[1\bar{1}0]$. Thus, taking into account the symmetry of the substrate, the two domains represent opposite chirality regarding the entire structure of substrate and film.

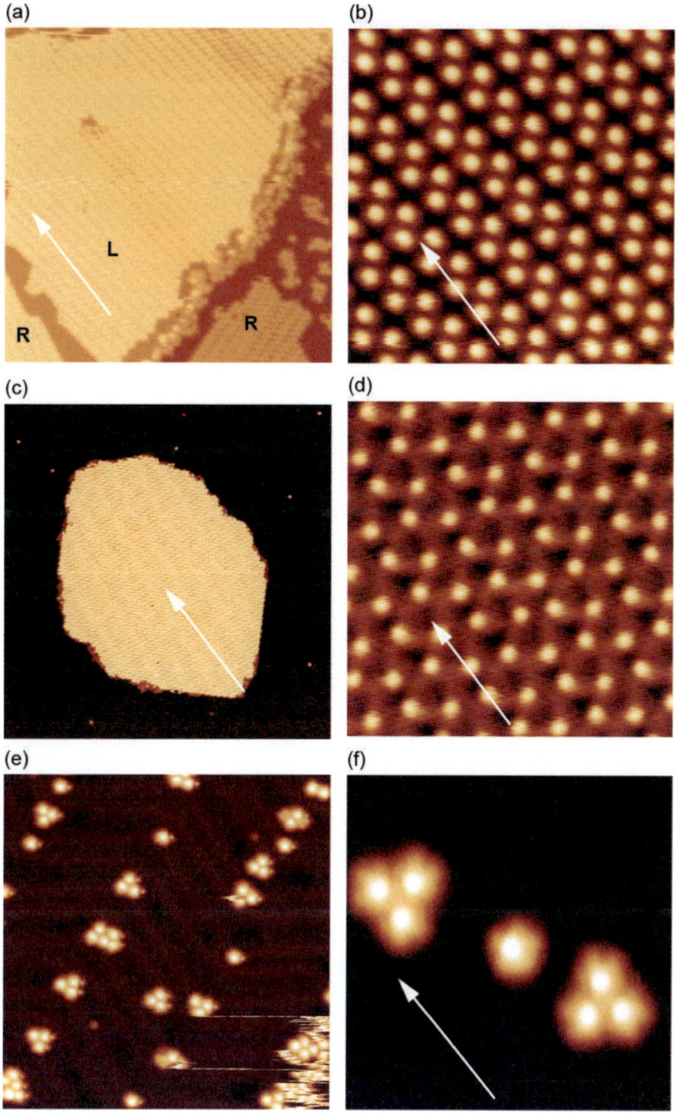

Figure 5.2: STM topographies of $Tb(thd)_3$ (a,b) on Cu(111), (c,d) on Ag(111), and (e,f) on Au(111). The image sizes are (a) 50×50 nm^2, (c) 180×180 nm^2, (e) 40×40 nm^2 and (b,d,f) 10×10 nm^2. The white arrows indicate the direction of $[1\bar{1}0]$ on the (111) surfaces. The labels R and L in (a) show the chirality of the film. Set point: (a,b) -0.8 V, 70 pA, (c) -0.8 V, 100 pA, (d) -1 V, 50 pA, (e,f) 1 V, 50 pA.

To investigate the superstructure of the molecular films, we superimposed a schematic model of Cu(111) or Ag(111) lattice on the STM topographies of the films with enhanced contrast, as shown in Figure 5.3(a,b). The analysis reveals that the molecules are assembled to films in a commensurate manner. Note that the adsorption sites of the molecules with respect to the lateral position (on top, three-fold hollow site, etc.) are not determined in our model since we were unable to obtain atomically resolved images of the substrates and molecules at the same time. The unit cells can nevertheless be represented by using vector $\boldsymbol{s}$ and $\boldsymbol{t}$ ($\boldsymbol{s}'$ and $\boldsymbol{t}'$) for the lattice vector of Cu(111) (Ag(111)) as follows:

$$\begin{pmatrix} \boldsymbol{A} \\ \boldsymbol{B} \end{pmatrix} = \begin{pmatrix} 4 & 1 \\ -5 & 7 \end{pmatrix} \begin{pmatrix} \boldsymbol{s} \\ \boldsymbol{t} \end{pmatrix} \tag{5.1}$$

for films on Cu(111), and

$$\begin{pmatrix} \boldsymbol{A}' \\ \boldsymbol{B}' \end{pmatrix} = \begin{pmatrix} 9 & 0 \\ -5 & 10 \end{pmatrix} \begin{pmatrix} \boldsymbol{s}' \\ \boldsymbol{t}' \end{pmatrix} \tag{5.2}$$

for films on Ag(111). The complicated structure of the molecular films results from the competition between intermolecular interactions and molecule-substrate interactions. The formation of these commensurate lattices implies that the molecules are transferred onto substrates without decomposition and a relatively strong molecule-substrate interaction is maintained.

5.1.3 *dI/dV* spectra and maps

We took dI/dV spectra on a molecule in a trimer on Au(111) to determine the local density of states (DOS) of the adsorbed molecules. The spectra on a molecule (black line) and on the Au substrate (blue line) are shown in Figure 5.4(a). The most prominent features in the spectra are marked by arrows at 350 mV and 650 mV. The peak at about −400 mV on Au(111) is caused by the surface state of gold [119].

To check the particular distribution of the DOS of the molecules, we performed dI/dV mapping on the trimer. The dI/dV maps at bias voltages of −550 mV, 100 mV, 350 mV and 650 mV, together with their corresponding topographies, are shown in Figure 5.4(b). The corrugation of the DOS on the substrate represents the standing waves caused by impurity scattering of the electrons of the surface state [120]. At 350 mV and 650 mV, where the maxima in the dI/dV spectrum were found, the DOS is localized on the molecules. Therefore the corresponding shapes of dI/dV maps present the molecular orbitals distinctly, which significantly differ from the topography, for instance, the center of the molecules is bright in the topography, while it is dark in the dI/dV map. At the voltages other than 350 mV and 650 mV, the shapes of the local DOS are more similar to the topographic images. This implies that the topographic images more likely represent the geometrical height of the molecules.

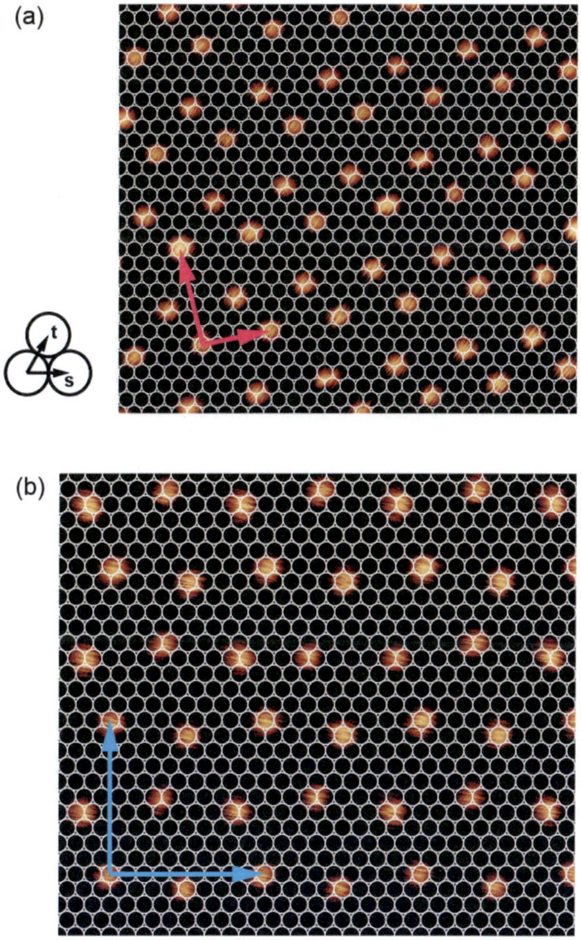

Figure 5.3: STM topographies of Tb(thd)$_3$ films (a) on Cu(111) and (b) on Ag(111) with the model of the substrate lattices. The white circles represent Cu or Ag atoms. The distance of nearest neighbouring Cu (Ag) atoms of substrate is 255 pm (288 pm) in this model. The vector $\boldsymbol{s}$ and $\boldsymbol{t}$ shown in left lower of (a) represents the lattice vectors of the substrate. The unit vectors of films are shown as red or blue arrows in each panel.

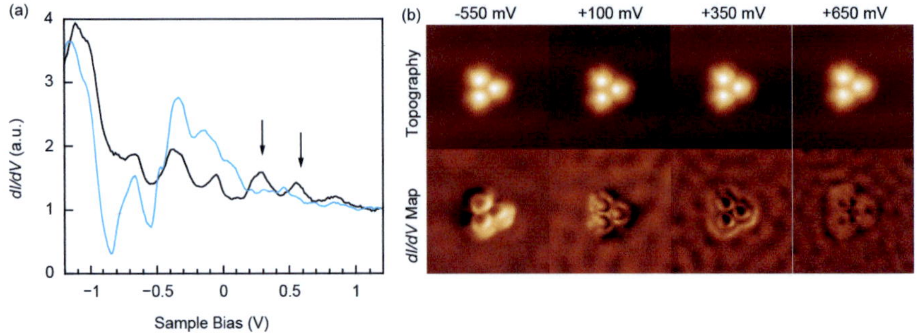

Figure 5.4: (a) dI/dV spectra of a molecule in a trimer (black line) and on the Au substrate (blue line). (b) Comparison of STM topographies and dI/dV maps under the same indicated voltages.

Unfortunately, our dI/dV spectra and maps do not present any feature related to spin excitation or other magnetic signals. Thus the magnetism of Tb(thd)$_3$ molecules is not evidenced. The absence of inelastic excitations can be ascribed to the shielding effect contributed by outer sub-shells on 4f electrons, which hence is decoupled from tunneling electrons. In parallel, the Kondo state of the Tb(thd)$_3$ molecule can be annihilated by charge transfer from the metal substrates, which can quench the magnetic momentum [88, 121].

5.2 [Tb(tmhd)$_3$]$_2$bpm

Benefiting from the interaction between two ions through the bridging ligand, the binuclear lanthanide complexes exhibit huge potential to realize quantum gates (qugates) [122], which is an attractive advantage in comparison to mono-nuclear complexes. A Kondo resonance has been observed on a single binuclear complex in a break junction [123], or even locally on the bridging ligand of a binuclear complex in the junction of our STM [124]. As one species of binuclear lanthanide complexes, [Ln(tmhd)$_3$]$_2$bpm polycrystals have been evidenced to exhibit single molecule magnet behaviour at low temperatures [109]. Therefore, it makes sense to study these molecules in lower dimensions. In this project, [Tb(tmhd)$_3$]$_2$bpm molecules were deposited onto Au(111) surface (see Section 3.2). The topography and differential conductance spectra of the assembled molecules were investigated by STM. All the experimental measurements were performed at 5 K.

The [Tb(tmhd)$_3$]$_2$bpm molecules were synthesised by our cooperators from the group of Prof. Mario Ruben in Karlsruhe [109]. A ball-stick model of a molecule is shown in Figure 5.5, where two terbium ions are bridged by a 2,2'-bipyrimidine (bpm) ligand, six additional

2,2',6,6'-tetramethyl-2,4-heptanedionate (tmhd) ligands are employed as peripheral ligands to support the Tb ions as well as to complete the charge-neutral compound.

Figure 5.5: A ball-stick model of a $[Tb(tmhd)_3]_2bpm$ molecule. Note that hydrogen atoms and methyl groups are omitted fro clarity.

5.2.1 Configurations of $[Tb(tmhd)_3]_2bpm$ molecules deposited on Au(111)

The configuration of $[Tb(tmhd)_3]_2bpm$ molecules deposited on Au(111) surface is displayed in Figure 5.6(a). Most of the molecules are self-assembled in large scale well-organized monolayer films including some decomposed ones. Some molecules also form individual clusters. Furthermore, ribbons consisting of two rows of molecules were observed. The typical herringbone reconstruction was visible on the bare Au(111) surface. A line profile of this topography is presented in Figure 5.6(b). Thus, the approximate heights of molecules can be derived: The heights are 360 pm and 280 pm for molecules in the clusters and in the film, respectively. Most likely, the compounds in the left bottom corner of Figure 5.6(a) are decomposed molecules, e.g. $Tb(thmd)_3$.

To investigate the well-organized molecular film, the topography of the same area was scanned with higher spacial resolution using various set points (see Figure 5.7). The values of the set point, especially the bias voltage affect the topography significantly, since the local density of states on the molecules is very sensitive to the bias voltage. However, the construction of the film was unchanged. The regular molecular assembly presents a six-fold symmetry. One of the striking hexagonal patterns is highlighted by a white hexagon in Figure 5.7(c). In the center of this hexagon, there is a protrusion with a "C"-shape opening.

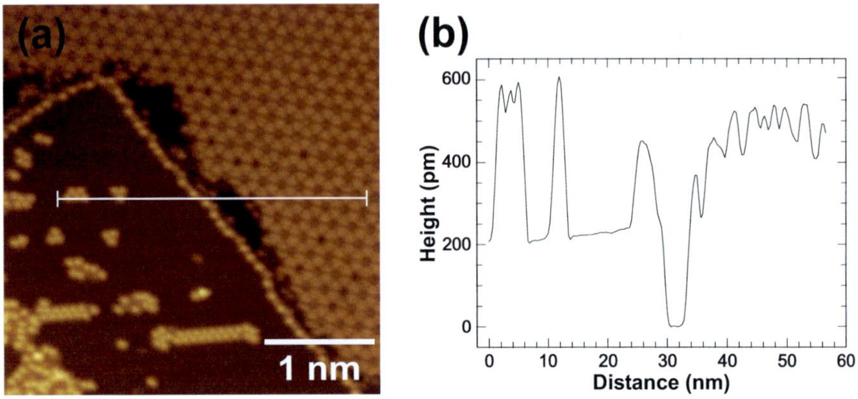

Figure 5.6: (a) STM topography of $[Tb(tmhd)_3]_2bpm$ molecules on Au(111) surface, set point: 2.5 V, 10 pA. (b) Line profile along the white line in (a).

In different hexagonal patterns, the "C"-shape opening of central protrusions points to six different directions. Moreover, in some patters, the central protrusion disappears. The distance between the nearest centres of hexagonal patterns is $\approx$ 3.4 nm. There are other six brighter dots located in the corners of each hexagonal pattern. Every three closest dots from three neighbouring patterns form a triangle. The distance between the centres of two closest dots is $\approx$ 1.0 nm. In Figure 5.7(c), one triangle is illustrated by a black triangular frame. In this image, the dots are elongated. Remarkably, the triangles present a coherent chirality. According to the dimensions of the substances in the images, we deduce that each bright dot refers to one $[Tb(tmhd)_3]_2bpm$ molecule. The chiral triangles imply that every three molecules interlock with each other thence form a well-organized film. However, the underlying origin of the hexagonal structure remains unknown.

Besides in the film described above, similar "C"-shape opening was also observed on molecules in individual cluster (see Figure 5.8(a)). Again, the opening presents different orientations. The origin of this opening has not been understood completely. The differential conductance dI/dV map taken on same cluster reveals the shape of the unoccupied molecular orbitals at 3 V (see Figure 5.8(b)).

Figure 5.7: STM images of the same area on the molecular film for different set point: 2 V, 50 pA for (a); 2 V, 40 pA for (b) and 3 V, 40 pA for (c). The whit hexagon and black triangle in (c) indicate a hexagonal pattern and a group of three interlocked molecules, respectively.

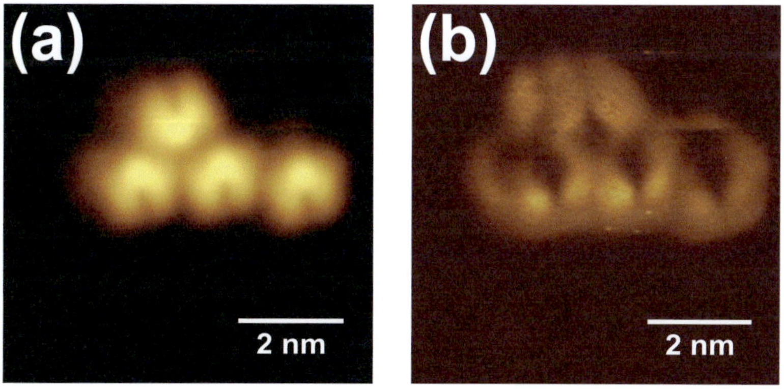

Figure 5.8: STM image (a) and corresponding dI/dV map (b) of an individual molecular cluster. Set point: 3 V, 36 pA. Lock-in modulation amplitude: 20 mV, frequency: 487 Hz.

5.2.2 *dI/dV* spectra

Figure 5.9 shows a comparison of the scanning tunneling spectra taken on a well-organized molecular film and on the bare Au(111) surface as a background. The latter one is identified by the peak at about −400 mV resulting from the surface state of gold [119]. In contrast to the background, a peak appeared approximately at 600 mV in the spectrum of the molecular film, which refers to an unoccupied molecular orbital above the Fermi level of $[Tb(tmhd)_3]_2bpm$. However, similar to the case of $Tb(thd)_3$ molecules on metallic surfaces, no inelastic spin excitation or Kondo resonance was observed on $[Tb(tmhd)_3]_2bpm$ molecules on Au(111).

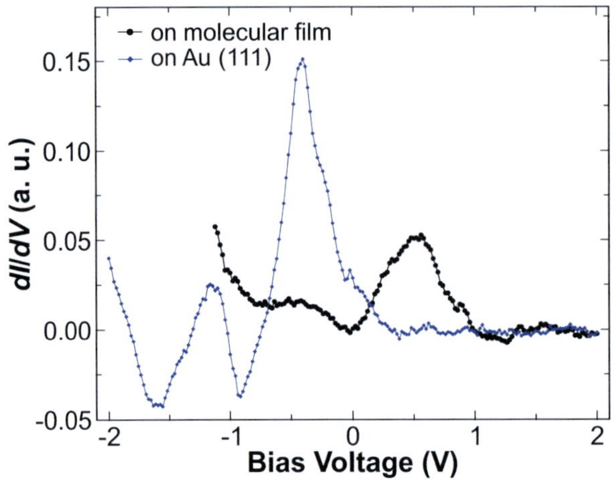

Figure 5.9: Comparison of dI/dV spectra taken on well-organized molecular film (black) and on the bare Au(111) substrate (blue). Lock-in modulation amplitude: 20 mV, frequency: 487 Hz.

5.3 Summary

We fabricated not only $Tb(thd)_3$ molecular assemblies on various metal surfaces but also $[Tb(tmhd)_3]_2bpm$ molecular assemblies on Au(111). By STM and STS at 5 K, the topography and spectroscopy of these assemblies were investigated. It is evidenced that $Tb(thd)_3$ and $[Tb(tmhd)_3]_2bpm$ complexes can be transferred onto metallic substrates by sublimation with few decomposition or surface contamination. For $Tb(thd)_3$, we showed that well-ordered monolayer films of lanthanide molecules are formed on Cu(111) and Ag(111), however isolated molecules and clusters stuck at elbow sites on Au(111). While $[Tb(tmhd)_3]_2bpm$ molecules form films with hexagonal structure on Au(111). Additionally, the dI/dV spectra taken on molecules present peaks. The dI/dV maps at different bias give us some information about the orbital distribution of the molecular assemblies. However, no feature related to a magnetic signal was observed, since the inelastic excitation can be suppressed by the shielding effect on the $4f$ electrons, while the Kondo effect can be annihilated by charge transfer from the substrates. Due to the chemical similarity of all rare earth elements, we speculate that similar complexes with different lanthanide cores can be sublimed as well. Thus, with this work, we have broadened the molecular species available for the study of rare-earth molecular magnetism. As a future outlook, similar to the results presented in the last chapter, it should also be possible to detect the spin excitations of proper magnetic molecules with $4f$ ions with STM at low temperatures on single-molecule level.

6 Spin crossover molecules on ferromagnetic substrates

As promising candidates to build organic electronic switches at nano-scale, spin crossover (SCO) molecules have attracted a great quantity of attention in research, since they display a transition between a low-spin (LS) and a high-spin (HS) state upon various external stimuli such as temperature, light, electric and magnetic field or charge flow. As discussed in Section 2.4, the mechanism of SCO phenomenon can be explained by the switching of the electron configuration of the magnetic ion, which in turn is caused by a change of ligand field.

The recent work of our group has shown that Fe(1,10-phenanthroline)$_2$(NCS)$_2$ (Fe-phen for short) complexes can be deposited onto various surfaces and consequently exhibits a coexistence of LS and HS states [37, 125]. In contrast to the Fe-phen molecules deposited on metallic surfaces (Cu(111), Cu(100), Au(111), Ag(111) and Co/Cu(111)) displaying frozen spin states, a Fe-phen molecule on Cu_2N/Cu(100) could be switched between LS and HS states by the tunneling current in scanning tunneling microscopy (STM). In addition, the ferromagnetic coupling between individual Fe-phen molecules at HS state and the thin film of Co below them has been evidenced [126]. However, Fe-phen still cannot be switched by a tunneling current on Co layer.

In this chapter, two hybrid systems composed of SCO molecules and a ferromagnetic substrate will be discussed. In the first system, the spin states of Fe-phen molecules on a ferromagnetic Fe_2N/Cu(100) layer were switched by tunneling currents at 5 K, which is similar to the case on Cu_2N/Cu(100). In the other system, the magnetization of LSMO covered by Fe-phen molecules with different thickness were examined by MOKE at temperature above $-120\,°C$. The magnetic properties of both system can be potentially manipulated by altering the exchange interaction between the two components due to the spin crossover. The Fe-phen molecules were synthesised by our cooperators from the group of Prof. Eric Beaurepaire in Strasbourg, France.

6.1 Spin crossover on Fe-phen molecules

Fe-phen has been noticed as a prototypical SCO molecule since the 1960s [127, 128]. As sketched in Figure 6.1, in this molecule, an iron ion at an oxidation state of +2 is surrounded by two NCS groups and two phenanthroline (phen) groups. The Fe^{2+} ion performs a thermal SCO (see Section 2.4) transition at 175 K from a HS state to a LS state as temperature decreases [128, 129], which results from a molecular deformation due to the shortening of the lengths of Fe-N bounds as well as the alternation of the angles between them [130]. In detail, Fe-phen at the HS state has a lower ligand field, such that the electrons obey Hund's rule. Therefore, the total spin is equal to 2, and the molecules are paramagnetic. However, when ligand field is large enough to overcome Hund's rule after the molecular deformation, Fe-phen shows a non-magnetic LS state at low temperatures.

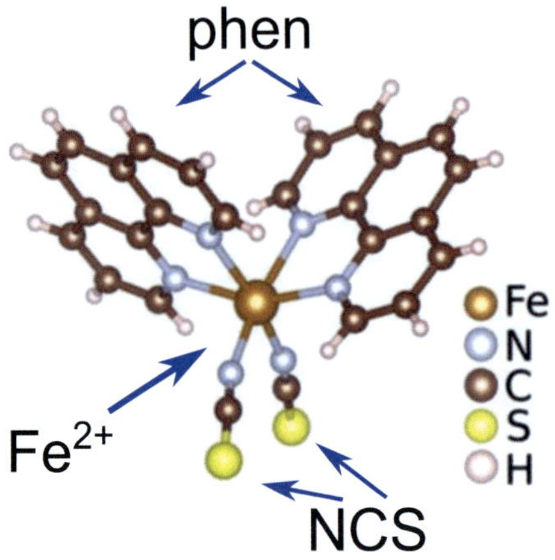

Figure 6.1: A ball-stick model of a Fe(1,10-phenanthroline)$_2$(NCS)$_2$ (Fe-phen) molecule.

6.1.1 Review of Fe-phen molecules on various substrates

Fe-phen molecules can be sublimed in UHV and deposited onto various surfaces [125]. In this section, the characteristics of Fe-phen molecules observed on different substrates will be reviewed.

The coexistance of two spin states

According to the previous study of Fe-phen with STM in our group, molecules anchor with NSC groups to the metal, while phen groups are exposed to the STM tip [37]. The STM images in Figure 6.2 reveal that there are two species of molecules with evidently different phen-phen distances on three metal surfaces. However, a Kondo resonance was only found on one species [37, 125], which should be the HS state. The Kondo impurity sits exactly in the middle of molecule at the position of the Fe ion, while, the other species in LS state never show a Kondo resonance. The identical behaviour happens for Fe-phen on Cu(100), Cu(111), Au(111) and Cu_2N/Cu(100). Furthermore, an X-ray magnetic circular dichroism (XMCD) study also indicated the coexistence of HS and LS species [37], since the spectrum is a superposition of two edges in the X-ray absorption spectrum of Fe-phen on Cu(100), which correspond to the two spin states, respectively. All of these observations demonstrate the coexistence of HS and LS spin state for Fe-phen molecules on various surfaces at low temperature.

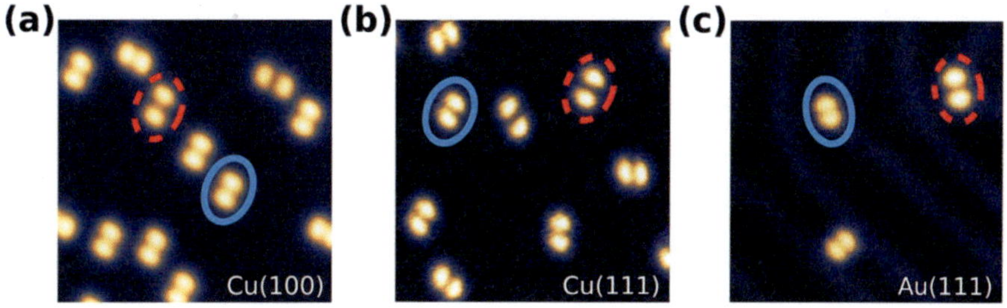

Figure 6.2: Topographies of submonolayer Fe-phen coverage on (a) Cu(100), (b) Cu(111), and (c) Au(111). On each topography, a HS (LS) Fe-phen molecule is encircled in dashed red (solid blue) for clarity. Image sizes are 14.1 nm×14.1 nm. This figure is reprinted with permission from Ref. [125].

Swithing Fe-phen on Cu_2N insulating layer

Actually, the spin states are determined by the adsorption site. For example, on Cu(100), when the Fe ion sits on a hollow site, Fe-phen always shows a HS state, while the Fe ion on bridge positions shows a LS state [131]. Unfortunately, one cannot switch the spin state of Fe-phen on metal, because the chemical bond between the Fe-phen and the metal is too strong.

To switch the spin states, we need a passivated surface to offer a weaker interaction. A copper nitride (Cu_2N) monolayer on Cu(100) is an excellent choice for that. It turned out Fe-phen can be switched back and forth between HS and LS state by the tunneling current [37].

Exchange coupling between Fe-phen and Co layer on Cu(111)

Recently, our group and our cooperators have investigated the exchange coupling between Fe-phen and a two monolayer Co below it [126]. HS species and possibly LS species were observed on the surface. Ferromagnetic coupling between HS Fe-phen and Co was confirmed both by theoretical calculation of density of states and by spin-polarized STM. However, Fe-phen still cannot be switched on the Co layer, which is consistent with the conditions on other metal surfaces.

6.2 Fe-phen molecules on Fe_2N

The motivation of this project is to combine the switching and coupling in one system. In that case, there will be a possibility to manipulate the magnetic properties of a hybrid system.

To guarantee the switching, we need a passivated surface. At the same time, the surface should be magnetically ordered for exchange coupling. So, we had to look for a particular surface to match our requirements. Luckily, recently, iron nitride (Fe_2N) on Cu(100) was found as a ferromagnetic layer [132, 133]. Similar to Cu_2N, it also offers a passivated surface. The method to prepare a monolayer of Fe_2N on Cu(100) has been introduced in Section 3.2. Fe-phen molecules were deposited to Fe_2N/Cu(100) by sublimation at 170 °C. The sample was measured at 5 K.

The quality of the Fe_2N layer is illustrated in Figure 3.2. The characteristic reconstruction of the Fe_2N layer can be recognized on the atomically resolved image. After depositing Fe-phen, we could find individual molecules as shown in Figure 6.3. However, unlike on Cu_2N, or metallic surfaces, Fe-phen on Fe_2N only shows one species at 4 K, since the molecules only present one type of topography.

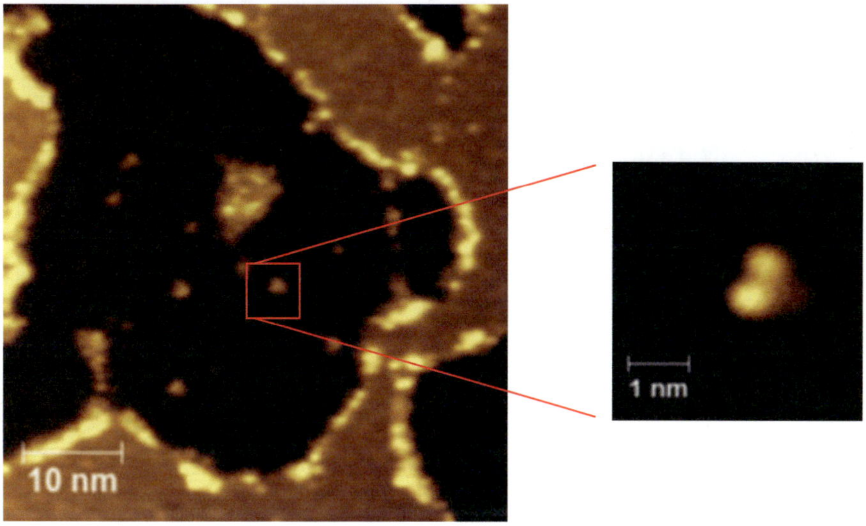

Figure 6.3: STM images of individual Fe-phen molecules on $Fe_2N/Cu(100)$ at 4 K. Set point: 1 V, 20 pA.

6.2.1 Switching spin states of Fe-phen on Fe_2N

Unlike the case of HS Fe-phen on other surfaces, a Kondo resonance was not observed in the differential conductance dI/dV spectrum of Fe-phen on Fe_2N (see Figure 6.4), which is expected since the Kondo state is prohibited on the ferromagnetic Fe_2N layer. However, a hysteresis was observed when sweeping an $I(V)$ cyclic curve on the center of Fe-phen as shown in Figure 6.5, which indicates the switching between high and low conductance. The abrupt changes of conductance occur approximately at ± 1.3 V during alternating the bias voltage with a constant tip-molecule distance. If we can treat this behaviour the same as for Fe-phen on Cu_2N [37], most likely, in our case of Fe-phen on Fe_2N, the high conductance also corresponds to the HS state, while the low conductance indicates the LS state. So, the goal to switch the spin states of Fe-phen on Fe_2N was achieved.

As a further confirmation, we tested reversible switching by bias pulses (see Figure 6.6). To do that, the STM tip was moved to the center of the molecule with an original set point of 10 mV, 5 pA. This tip-sample distance was kept, when 3 second-wide rectangular voltage pulses of either 1.8 V or -2.2 V were added to the background bias voltage of 10 mV after each break of about 5 s. Simultaneously, the response of the tunneling current was recorded as a function of time. We found that there are two levels in the tunneling current (indicated by red and blue stripes in the lower panel of Figure 6.6) corresponding to the HS state

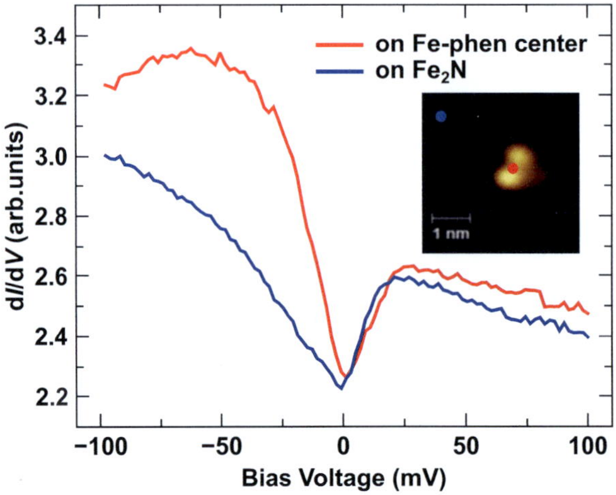

Figure 6.4: Comparision of dI/dV spectra taken on the center of an Fe-phen molecule and on bare Fe_2N surface with a lock-in modulation bias of 4 mV and a frequency of 763 Hz. The red and blue dots in the inset indicate the positions of the tip where the spectra were taken. Set point: 100 mV, 500 pA.

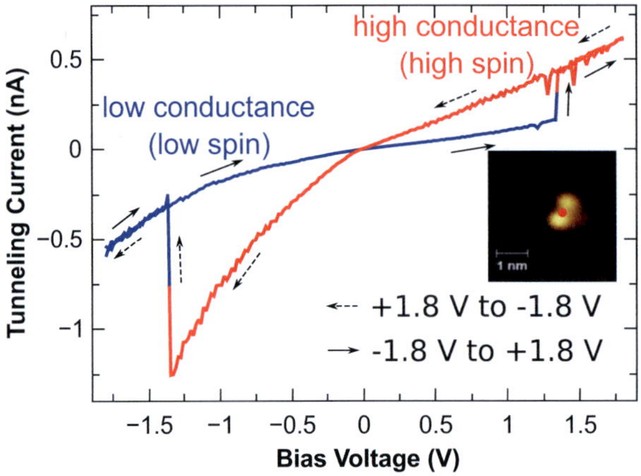

Figure 6.5: A hysteretic switching in an $I(V)$ curve taken on the center of an Fe-phen molecule at 4 K. The red dot in the inset indicates the positions of the tip where the curve was taken. Set point: 1.5 V, 500 pA.

and the LS state, respectively. Notice that the LS state can be switched to the HS state by voltage with either polarity. But the HS state can be switched to the LS state only by negative pulse. Furthermore switching is not guaranteed. We believe that the HS state has lower energy, which also explains why only one species was observed in the topography, which corresponds to the ground state.

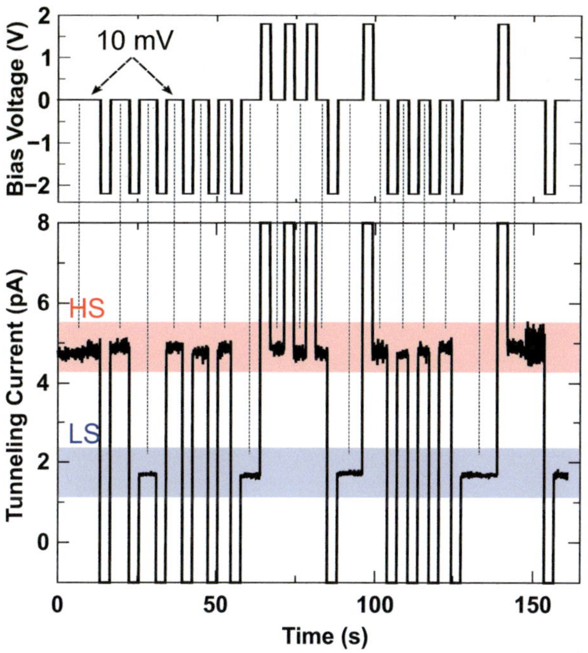

Figure 6.6: Reversible switching the spin state of the Fe^{2+} ion by bias pulses. Upper panel: the mixed bias voltage applied on the sample as a function of time. Lower panel: the response of the tunneling current to the oscillating voltage. The red (blue) stripe indicates that the high (low) spin state. During the measurement, the tip-molecule distance was kept at the initial set point: 10 mV, 5 pA.

Similar to Cu_2N, we also found that the switching from the HS state to the LS state is not instantaneous for Fe-phen molecules on Fe_2N, since a drop in current with a delay of half second after the pulse was observed (see Figure 6.7(a) and (b)). When starting to add the negative pulse of −2.2 V, the tunneling current immediately shot to ≈−4 nA, then abruptly decreased to ≈−1 nA after 0.5 s as indicated by the arrows in Figure 6.7(b), although the applied negative bias voltage was not changed in 3 s. In contrast, from the LS state to the HS state, it is a fast switch, as the tunneling current stayed at about 1 nA during the whole span of the 1.8 V positive pulse (see Figure 6.7(c) and (d)). To explain the switching mechanism, we can use the same model as for the case of Fe-phen on Cu_2N introduced in Ref. [37]. The process from the LS state to the HS state can be explained by heating effect of tunneling electrons, which induces direct and fast switching, while the switching from the HS state to the LS state corresponds to a single-electron process involving an intermediate step. In detail, the Fe^{2+} ion in the HS state is first brought to an excited state by one tunneling electron, thereupon relaxes to the LS state. The low transition probability of this single-electron process explains the delay of the switching.

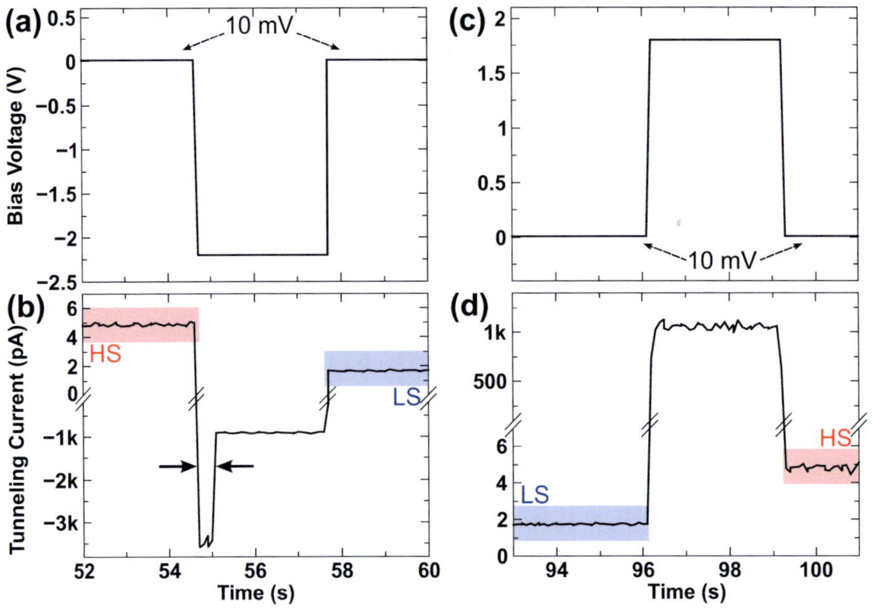

Figure 6.7: The trace of bias voltage when a negative (a) or a positive (c) pulse was applied, and corresonding current trace (b and d), respectively. The arrows in (b) indicate the time delay of the switch from the HS state to the LS state occured after the negative pulse had been applied.

6.2.2 Partial summary

In summary, Fe-phen molecules were deposited on Fe_2N/Cu(100). There is only one ground state for Fe-phen, which means there is no coexistence of the two species. But Fe-phen can be switched between two states with tunneling current. As an outlook, we can investigate the exchange coupling between Fe-phen in the HS state and ferromagnetic Fe_2N layer by using a spin-polarized tip. Eventually, there is the possibility to manipulate the magnetic properties of Fe_2N by switching the states of Fe-phen, for instance with light.

6.3 Fe-phen molecules on LSMO

Besides on monolayer of Fe_2N at 4 K, we also investigated Fe-phen molecules on half-metal lanthanum strontium manganite (LSMO) at higher temperatures. As a prototypical perovskite manganite, LSMO has been widely studied since 1950s [134], since it exhibits diverse magnetic properties. For instance, with optimal doping, the bulk of $La_{0.7}Sr_{0.3}MnO_3$ is ferromagnetic with a Curie temperature as high as 370 K. In parallel, as mentioned earlier in this chapter, the Fe-phen molecules stay in the HS state at temperatures above 175 K. Moreover, the magnetic coupling between the Fe^{2+} ions in the HS state and the Co layer below has been evidenced [126]. Regarding these effects, in the presence of these magnetic molecules on the surface, the LSMO film should also magnetically couple to them above 175 K. Thus, there is a chance for us to evidence the coupling by using our longitudinal magneto-optic Kerr effect (MOKE) set-up (see Section 3.1.2). A straightforward idea is to trace the magnetic hysteresis loops of LSMO with and without Fe-phen molecules on top as a function of temperature. A difference in the temperature dependent evolution of the loops could be observed due to the appearance of the exchange coupling in the system of LSMO with Fe-phen. Furthermore, an abrupt change of the loop shape might result from the emerging or the vanishing of the exchange coupling when the SCO of the Fe-phen molecules occurs after the transition temperature. If this proposal works out, there will be a possibility to manipulate the magnetic property of the system by switching the spin state of Fe-phen molecules at relatively high temperatures.

6.3.1 Temperature dependent magnetizations

The magnetic property of the 10 nm-thick $La_{0.7}Sr_{0.3}MnO_3$ (LSMO) film epitaxially grown on $SrTiO_3(100)$ (STO) was measured at first as a reference. Subsequently, after a deposition of 5 min, the sample of the LSMO film with Fe-phen molecules on top was measured. To ensure that LSMO was completely covered by Fe-phen molecules, additional molecules were deposited onto the sample for 25 min more. Then the magnetic properties were measured, again. The details about the substrate treatment and molecule deposition have been introduced in Section 3.2.

Before measuring the in-plane magnetization, the head of the manipulator where the samples sat was cooled down with liquid nitrogen below $-120\,°C$. Afterwards, the magnetic hysteresis loops at various temperatures were measured with longitudinal MOKE during the slow warming up process. The temperature was monitored by a thermocouple attached on the head of the manipulator rather than on the sample directly, for which the error is estimated to be below 5 °C.

As revealed by the hysteresis loops shown in Figure 6.8, all three samples experienced phase transition from ferromagnetic state to paramagnetic state as temperature increases. The square-like hysteresis loops at temperature below −100 °C imply that this in-plane direction is not very hard to magnetize. Unfortunately, we could not calibrate the magnetic field reaching the samples, but the relative strength of the field was indicated by the current applied in the electromagnet. The coercive field of LSMO at low temperatures should be in the range from 5 Oe to 10 Oe according to literature [135]. The opening of the hysteresis vanished at temperatures below −10 °C, which refers to the Curie temperature of the LSMO film. Note that instead of a single loop, a double-loop behaviour started to appear in hysteresis loops as the temperature approached to the Curie temperature. This phenomenon implies that two phases with different coerecivities coexist in our LSMO film at particular temperatures or the samples have two easy axes.

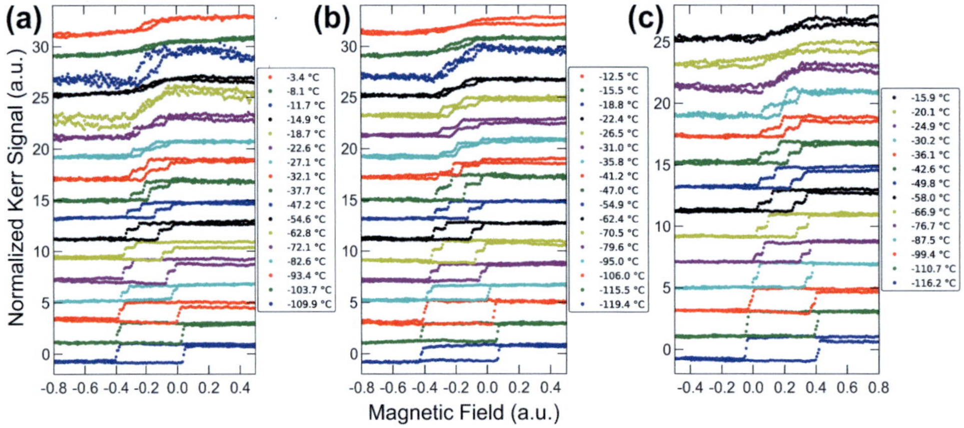

Figure 6.8: Magnetic hysteresis loops measured by MOKE on samples of pure 10 nm-thick LSMO film (a), with Fe-phen molecules deposited on top for 5 min (b) and subsequently another 25 min (c) at various temperatures as indicated. The magnetic field generated from an electromagnet was applied in the in-plane direction of the samples. The loops are vertically shifted for clarity. The magnetization of each loop is normalized by scaling the Kerr signal under the positive (negative) maximum field to 1 (−1).

To check the exchange coupling between the LSMO film and the Fe-phen molecules on it, the temperature dependent coercive field H_c and exchange bias of pure LSMO and samples with Fe-phen on the surface are compared (see Figure 6.9). H_c and the exchange bias extracted from loops shown in Figure 6.8 refer to the half-width of the opening and the lateral shift of the center with respect to zero magnetic field for each loop, respectively. As shown in Figure 6.9(a), H_c linearly decreased to zero as temperature increased to the Curie temperature for all of the three samples. However, there is no difference between the sample with and without Fe-phen molecules. Regarding the temperature dependent exchange biases of three samples shown in Figure 6.9(b), despite there is a disparity between the exchange bias of the sample with Fe-phen deposition of 30 min, we ascribe the non-zero values to the remnant field of the electromagnet, since they are constant during the whole temperature range of the measurement. Consequently, within the resolution of our longitudinal MOKE set-up, no evidence was obtained to support exchange coupling between the LSMO film and Fe-phen molecules on it.

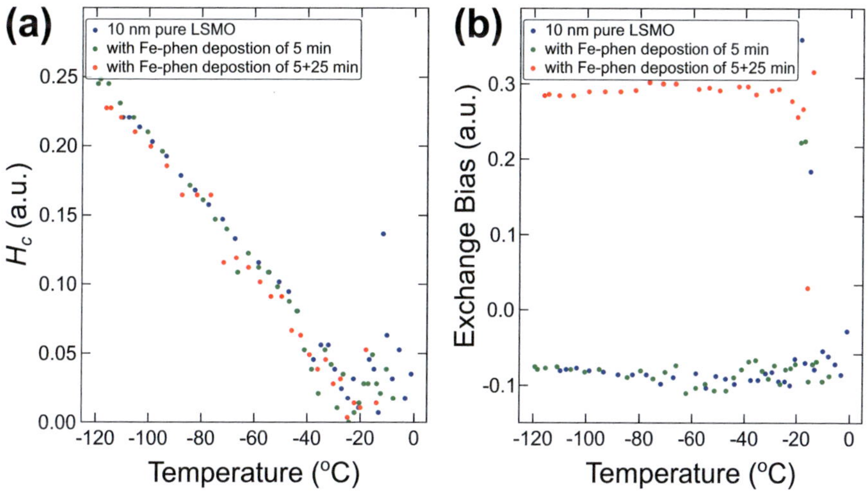

Figure 6.9: Comparison of temperature dependent coercive field H_c (a) and exchange bias (b) of pure LSMO film, sample with Fe-phen deposition of 5 min and sample with another depostion of 25 min. For each loop, H_c refers to the half-width of the opening. Exchange bias refers to the lateral shift of the center with respect to zero magnetic field.

6.3.2 Outlook

In spite of the fact that the exchange interaction between Fe-phen molecules and ferromagnetic LSMO substrates were not evidenced by our brief MOKE experiments, we are inspired to investigate the magnetic properties of the same system with XMCD in a synchrotron, where the magnetic moments of Fe ion and Mn ion can be detected separately.

7 Conclusion and prospect

Owing to the requirement of high density magnetic data storage and the pursuit of quantum computing, we were attracted by the novel properties and promising potential applications of single magnetic molecules. This dissertation was performed to understand the behaviour of a series of single magnetic molecules, the exchange coupling between them and the substrates underneath. Scanning tunneling microscopy (STM) at low temperatures down to 30 mK with magnetic fields up to 5 T as well as a longitudinal magneto-optic Kerr effect (MOKE) set-up working above $-120\,^{\circ}$C were employed in our experiments. The adsorption geometries of different single magnetic molecules deposited on various surfaces were revealed by the STM topographies. The magnetic properties, or more precisely speaking, the spin states of molecules at low temperatures, were investigated by scanning tunneling spectroscopy (STS). Additionally, the temperature dependent magnetization of one system was studied with MOKE. The main findings for each category of molecules are summarised below in detail.

M-bis(2,2,6,6-tetramethyl-3, 5-heptanedionato) (M = Co, Ni and Cu) (for short $M(thd)_2$) and manganese phthalocyanine (MnPc) were studied as two examples of magnetic molecules containing single 3*d* ions. A discontinuous change of the ligand fields around the core ions in $M(thd)_2$ molecules upon formation of contact was evidenced. This consequently changes the order of the spin states in the energy spectrum. In addition, a Kondo resonance was observed on $Co(thd)_2$ on $Cu_2N/Cu(100)$. The planar MnPc molecules lie flat on Cu(100) and Co/Cu(100). The molecules of the first three layers shack with the same in-plane orientation. A pronounced Kondo resonance ($T_K \approx 207\,K$) and spin-flip inelastic excitations were observed on MnPc deposited on Cu(100) and Co/Cu(100) respectively, which demonstrate that single MnPc molecules are magnetic.

In order to study the performance of single molecules containing $4f$ ions, the mononuclear lanthanide complex $Tb(thd)_3$ and the binuclear lanthanide complex $[Tb(tmhd)_3]_2bpm$ were sublimed from evaporators and deposited onto various metal substrates without decomposition or contaminations. Homogeneous monolayer films on a large scale with well-defined reconstructions and clusters consisting of several molecules were formed on different surfaces due to different molecule-substrate interactions. The orbitals of molecules under particular bias voltages were revealed by STS.

Typical spin crossover (SCO) molecules, Fe(1,10-phenanthroline)$_2$(NCS)$_2$ (Fe-phen for short) were deposited onto ferromagnetic Fe_2N/Cu(100) and lanthanum strontium manganite (LSMO) films to fabricate hybrid systems which potentially exhibit exchange coupling. The spin state of Fe-phen on Fe_2N/Cu(100) was switched between high-spin and low-spin by tunneling currents at low temperature, in spite of that only one ground state was observed. For the system of LSMO covered by Fe-phen molecules, MOKE revealed its temperature dependent magnetization in the range from -120 to $0\,°C$.

Considering the prospect and potential applications, the following suggestions may be helpful.

(i) The strong tuning on the spin states of M(thd)$_2$ by molecular deformation can be applied locally on individual molecules and has the potential to prepare superposition states in a way that avoids the use of microwave excitation in magnetic fields. To achieve the superposition states, we suggest to design systems with long coherence times of spins or use surface acoustic waves to induce molecular deformations.

(ii) The revealed adsorption geometries demonstrated how various single magnetic molecules were assembled on different surfaces as the consequences of the competition between molecule-molecule interaction and molecule-substrate interaction. The understanding of the interfaces between organic molecules and the underneath metallic substrates is crucial for designing organic spintronic devices.

(iii) Although no magnetic signal was detected on two lanthanide-based complexes in our experiment, we broadened the categories of single molecules containing $4f$ ions that can be transferred to surfaces and whose magnetic properties can be studies with STM at low temperatures. Similar complexes can be promising candidates for molecular magnets with single $4f$ ions, which carry larger magnetic moments and exhibit longer magnetization relaxation times.

(iv) With respect to the SCO effect of Fe-phen, a succeeding project may be to study the tunable exchange coupling between the Fe-phen molecules and the ferromagnetic Fe_2N layer with spin-polarized STM. Eventually, the magnetic properties of Fe_2N can be potentially manipulated by switching the states of Fe-phen, for instance with light.

(v) Ascribed to the complicated situation of the interface and our limited MOKE set-up, the coupling between Fe-phen molecules and ferromagnetic LSMO films was not evidenced in this dissertation. A plan of investigating the same system with X-ray magnetic circular dichroism (XMCD) in the synchrotron in Karlsruhe is currently being worked out.

Our investigations of individual molecules are crucial to understand the physics of single magnetic molecules. We anticipate that the results and analyses of this dissertation illustrate novel research objects and crucial progresses on molecular electronics and spintronics. There should be more achievements by solving the remaining questions and advancing the prospects in the future.

Abbreviations and symbols

The next list describes abbreviations and symbols which are used within the dissertation

Abbreviation

CFT	Crystal field theory
$Co(thd)_2$	Bis(2,2,6,6-tetramethyl-3, 5-heptanedionato) 3d cobalt(II)
$Cu(thd)_2$	Bis(2,2,6,6-tetramethyl-3, 5-heptanedionato) 3d copper(II)
DOS	Density of states
Fe-phen	$Fe(1,10\text{-phenanthroline})_2(NCS)_2$
HS	High-spin
IETS	Inelastic electron tunneling spectroscopy
LDOS	Local density of states
LFT	Ligand field theory
$Ln(thd)_3$	Tris(2,2,6,6-tetramethyl-3,5-heptanedionato)lanthanide(III)
$LnPc_2$	Bis(phthalocyaninato)lanthanide
LS	Low-spin
LSMO	Lanthanum strontium manganite
$M(thd)_2$	Bis(2,2,6,6-tetramethyl-3, 5-heptanedionato) 3d metal(II)
ML	Monolayer
MnPc	Manganese phthalocyanine
MOKE	Magneto-optic Kerr effect
$Ni(thd)_2$	Bis(2,2,6,6-tetramethyl-3, 5-heptanedionato) 3d nickel(II)
Pc	Phthalocyanine
SCO	Spin crossover

SIM	Single-ion magnet
SMM	Single magnetic molecule
SOC	Spin-orbit coupling
STM	Scanning tunneling microscope
STO	Strontium titanate
STS	Scanning tunneling spectroscopy
$Tb(thd)_3$	Tris(2,2,6,6-tetramethyl-3,5-heptanedionato)terbium(III)
UHV	Ultra-hight vacuum
XMCD	X-ray magnetic circular dichroism
ZFS	Zero field splitting

Symbol

$\mathbf{Y}_n^m(\theta,\phi)$	Spherical harmonics
$\mathscr{H}$	Hamiltonian

List of Figures

List of Tables

Bibliography

[1] B Bleaney and K W H Stevens. Paramagnetic resonance. *Reports on Progress in Physics*, 16(1):108–159, 1953.

[2] Roman Boča. Zero-field splitting in metal complexes. *Coordination Chemistry Reviews*, 248(9-10):757–815, 2004.

[3] Benjamin W. Heinrich, Lukas Braun, Jose I. Pascual, and Katharina J. Franke. Tuning the magnetic anisotropy of single molecules. *Nano Letters*, 15(6):4024–4028, 2015.

[4] J. Kondo. Resistance minimum in dilute magnetic alloys. *Prog. Theor. Phys.*, 32:37–49, 1964.

[5] J. S. Griffith and L. E. Orgel. Ligand-field theory. *Q. Rev. Chem. Soc.*, 11:381–393, 1957.

[6] Mildred S. Dresselhaus, Gene Dresselhaus, and Ado Jorio. *Group theory : application to the physics of condensed matter*. Springer, Berlin, 2008.

[7] K W H Stevens. Matrix elements and operator equivalents connected with the magnetic properties of rare earth ions. *Proceedings of the Physical Society. Section A*, 65(3):209–215, 1952.

[8] Brian G. Wybourne and William F. Meggers. Spectroscopic properties of rare earths, 1965.

[9] M. T. Hutchings. Point-charge calculations of energy levels of magnetic ions in crystalline electric fields. *Solid state physics*, 16:227–273, 1964.

[10] E. Bauer and M. Rotter. *Magnetism of complex metallic alloys: crystalline electric field effects*, pages 183–248. World Scientific, 2011.

[11] J. Bardeen. Tunnelling from a many-particle point of view. *Phys. Rev. Lett.*, 6:57–59, Jan 1961.

[12] J. Tersoff and D. R. Hamann. Theory and application for the scanning tunneling microscope. *Phys. Rev. Lett.*, 50:1998–2001, Jun 1983.

[13] C. Julian Chen. *Introduction to Scanning Tunneling Microscopy*. Oxford University Press, 1993.

[14] J. Lambe and R. C. Jaklevic. Molecular vibration spectra by inelastic electron tunneling. *Phys. Rev.*, 165:821–832, Jan 1968.

[15] J. Klein, A. Léger, M. Belin, D. Défourneau, and M. J. L. Sangster. Inelastic-electron-tunneling spectroscopy of metal-insulator-metal junctions. *Phys. Rev. B*, 7:2336–2348, Mar 1973.

[16] W. Meissner and B. Voigt. Messungen mit Hilfe von flüssigem Helium XI Widerstand der reinen Metalle in tiefen Temperaturen. *Annalen der Physik*, 399(8):892–936, 1930.

[17] W.J. de Haas, J. de Boer, and G.J. van dën Berg. The electrical resistance of gold, copper and lead at low temperatures. *Physica*, 1(7):1115–1124, 1934.

[18] P. W. Anderson. Localized magnetic states in metals. *Phys. Rev.*, 124:41–53, 1961.

[19] Kenneth G. Wilson. The renormalization group: Critical phenomena and the Kondo problem. *Rev. Mod. Phys.*, 47:773–840, Oct 1975.

[20] F. Patthey, J.-M. Imer, W.-D. Schneider, H. Beck, Y. Baer, and B. Delley. High-resolution photoemission study of the low-energy excitations in $4f$-electron systems. *Phys. Rev. B*, 42:8864–8881, Nov 1990.

[21] M. Garnier, K. Breuer, D. Purdie, M. Hengsberger, Y. Baer, and B. Delley. Applicability of the single impurity model to photoemission spectroscopy of heavy fermion Ce compounds. *Phys. Rev. Lett.*, 78:4127–4130, May 1997.

[22] D. C. Ralph and R. A. Buhrman. Kondo-assisted and resonant tunneling via a single charge trap: A realization of the Anderson model out of equilibrium. *Phys. Rev. Lett.*, 72:3401–3404, May 1994.

[23] Jiutao Li, Wolf-Dieter Schneider, Richard Berndt, and Bernard Delley. Kondo scattering observed at a single magnetic impurity. *Phys. Rev. Lett.*, 80:2893–2896, Mar 1998.

[24] Markus Ternes, Andreas J Heinrich, and Wolf-Dieter Schneider. Spectroscopic manifestations of the Kondo effect on single adatoms. *Journal of Physics: Condensed Matter*, 21(5):053001, 2009.

[25] Alexander Cyril Hewson. *The Kondo Problem to Heavy Fermions*. Cambridge Studies in Magnetism. Cambridge University Press, 1993.

[26] K. Nagaoka, T. Jamneala, M. Grobis, and M. F. Crommie. Temperature dependence of a single Kondo impurity. *Phys. Rev. Lett.*, 88:077205, Feb 2002.

[27] Alexander F Otte, Markus Ternes, Kirsten von Bergmann, Sebastian Loth, Harald Brune, Christopher P Lutz, Cyrus F Hirjibehedin, and Andreas J Heinrich. The role of magnetic anisotropy in the Kondo effect. *Nature Physics*, 4(11):847–850, 2008.

[28] D. Goldhaber-Gordon, Hadas Shtrikman, D. Mahalu, David Abusch-Magder, U. Meirav, and M. A. Kastner. Kondo effect in a single-electron transistor. *Nature*, 391:156, Jan 1998.

[29] T. A. Costi. Kondo effect in a magnetic field and the magnetoresistivity of Kondo alloys. *Phys. Rev. Lett.*, 85:1504–1507, Aug 2000.

[30] U. Fano. Effects of configuration interaction on intensities and phase shifts. *Phys. Rev.*, 124:1866–1878, Dec 1961.

[31] V. Madhavan, W. Chen, T. Jamneala, M. F. Crommie, and N. S. Wingreen. Tunneling into a single magnetic atom: Spectroscopic evidence of the Kondo resonance. *Science*, 280(5363):567–569, 1998.

[32] P. Weinberger. John Kerr and his effects found in 1877 and 1878. *Philosophical Magazine Letters*, 88(12):897–907, 2008.

[33] Masud Mansuripur. *The physical principles of magneto-optical recording*. Cambridge University Press, 1998.

[34] P. R. Cantwell, U. J. Gibson, D. A. Allwood, and H. A. M. Macleod. Optical coatings for improved contrast in longitudinal magneto-optic Kerr effect measurements. *Journal of Applied Physics*, 100(9):093910, 2006.

[35] L. Zhang, T. Miyamachi, T. Tomanić, R. Dehm, and W. Wulfhekel. A compact sub-Kelvin ultrahigh vacuum scanning tunneling microscope with high energy resolution and high stability. *Review of Scientific Instruments*, 82(10):103702, 2011.

[36] FM Leibsle, SS Dhesi, SD Barrett, and AW Robinson. Stm observations of Cu(100)-c(2×2)n surfaces: evidence for attractive interactions and an incommensurate c(2×2) structure. *Surface Science*, 317(3):309–320, 1994.

[37] Toshio Miyamachi, Manuel Gruber, Vincent Davesne, Martin Bowen, Samy Boukari, Loïc Joly, Fabrice Scheurer, Guillaume Rogez, Toyo Kazu Yamada, Philippe Ohresser, Eric Beaurepaire, and Wulf Wulfhekel. Robust spin crossover and memristance across a single molecule. *Nature Communications*, 3:938, Jul 2012.

[38] J. M. Gallego, S. Yu. Grachev, D. M. Borsa, D. O. Boerma, D. Écija, and R. Miranda. Mechanisms of epitaxial growth and magnetic properties of $\gamma'-Fe_4N(100)$ films on Cu(100). *Phys. Rev. B*, 70:115417, Sep 2004.

[39] J. Kirschner, H. Engelhard, and D. Hartung. An evaporation source for ion beam assisted deposition in ultrahigh vacuum. *Review of Scientific Instruments*, 73(11):3853–3860, 2002.

[40] A Gambardella, P Graziosi, I Bergenti, M Prezioso, D Pullini, S Milita, F Biscarini, and VA Dediu. Surface nanostructures in manganite films. *Scientific reports*, 4, 2014.

[41] G. Binnig, H. Rohrer, Ch. Gerber, and E. Weibel. Tunneling through a controllable vacuum gap. *Applied Physics Letters*, 40(2):178–180, 1982.

[42] G. Binnig, H. Rohrer, Ch. Gerber, and E. Weibel. Surface studies by scanning tunneling microscopy. *Phys. Rev. Lett.*, 49:57–61, Jul 1982.

[43] Hongjie Dai, Jason H. Hafner, Andrew G. Rinzler, Daniel T. Colbert, and Richard E. Smalley. Nanotubes as nanoprobes in scanning probe microscopy. *Nature*, 384(6605):147–150, Nov 1996.

[44] D. K. Biegelsen, F. A. Ponce, J. C. Tramontana, and S. M. Koch. Ion milled tips for scanning tunneling microscopy. *Applied Physics Letters*, 50(11):696–698, 1987.

[45] W. Hill Harman, T. David Harris, Danna E. Freedman, Henry Fong, Alicia Chang, Jeffrey D. Rinehart, Andrew Ozarowski, Moulay T. Sougrati, Fernande Grandjean, Gary J. Long, Jeffrey R. Long, and Christopher J. Chang. Slow magnetic relaxation in a family of trigonal pyramidal iron(II) pyrrolide complexes. *Journal of the American Chemical Society*, 132(51):18115–18126, 2010.

[46] Gavin A. Craig and Mark Murrie. 3d single-ion magnets. *Chem. Soc. Rev.*, 44:2135–2147, 2015.

[47] Lapo Bogani and Wolfgang Wernsdorfer. Molecular spintronics using single-molecule magnets. *Nature Materials*, 7(3):179–186, 2008.

[48] A. J. Heinrich, J. A. Gupta, C. P. Lutz, and D. M. Eigler. Single-atom spin-flip spectroscopy. *Science*, 306(5695):466–469, 2004.

[49] Cyrus F. Hirjibehedin, Chiung-Yuan Lin, Alexander F. Otte, Markus Ternes, Christopher P. Lutz, Barbara A. Jones, and Andreas J. Heinrich. Large magnetic anisotropy of a single atomic spin embedded in a surface molecular network. *Science*, 317(5842):1199–1203, 2007.

[50] Cyrus F. Hirjibehedin, Christopher P. Lutz, and Andreas J. Heinrich. Spin coupling in engineered atomic structures. *Science*, 312(5776):1021–1024, 2006.

[51] Xi Chen, Ying-Shuang Fu, Shuai-Hua Ji, Tong Zhang, Peng Cheng, Xu-Cun Ma, Xiao-Long Zou, Wen-Hui Duan, Jin-Feng Jia, and Qi-Kun Xue. Probing superexchange interaction in molecular magnets by spin-flip spectroscopy and microscopy. *Phys. Rev. Lett.*, 101:197208, Nov 2008.

[52] J. Fransson. Spin inelastic electron tunneling spectroscopy on local spin adsorbed on surface. *Nano Letters*, 9(6):2414–2417, 2009.

[53] Noriyuki Tsukahara, Ken-ichi Noto, Michiaki Ohara, Susumu Shiraki, Noriaki Takagi, Yasutaka Takata, Jun Miyawaki, Munetaka Taguchi, Ashish Chainani, Shik Shin, and Maki Kawai. Adsorption-induced switching of magnetic anisotropy in a single iron(II) phthalocyanine molecule on an oxidized Cu(110) surface. *Phys. Rev. Lett.*, 102:167203, Apr 2009.

[54] BW Heinrich, L Braun, JI Pascual, and KJ Franke. Protection of excited spin states by a superconducting energy gap. *Nature Physics*, 9(12):765–768, 2013.

[55] P. Wahl, L. Diekhöner, G. Wittich, L. Vitali, M. A. Schneider, and K. Kern. Kondo effect of molecular complexes at surfaces: Ligand control of the local spin coupling. *Phys. Rev. Lett.*, 95:166601, Oct 2005.

[56] C. Romeike, M. R. Wegewijs, W. Hofstetter, and H. Schoeller. Quantum-tunneling-induced Kondo effect in single molecular magnets. *Phys. Rev. Lett.*, 96:196601, May 2006.

[57] Gabriel González, Michael N. Leuenberger, and Eduardo R. Mucciolo. Kondo effect in single-molecule magnet transistors. *Phys. Rev. B*, 78:054445, Aug 2008.

[58] R. Hiraoka, E. Minamitani, R. Arafune, N. Tsukahara, S. Watanabe, M. Kawai, and N. Takagi. Single-molecule quantum dot as a Kondo simulator. *Nature Communications*, 8:16012, Jun 2017.

[59] Jiwoong Park, Abhay N. Pasupathy, Jonas I. Goldsmith, Connie Chang, Yuval Yaish, Jason R. Petta, Marie Rinkoski, James P. Sethna, Hector D. Abruna, Paul L. McEuen, and Daniel C. Ralph. Coulomb blockade and the Kondo effect in single-atom transistors. *Nature*, 417(6890):722–725, Jun 2002.

[60] Wenjie Liang, Matthew P. Shores, Marc Bockrath, Jeffrey R. Long, and Hongkun Park. Kondo resonance in a single-molecule transistor. *Nature*, 417(6890):725–729, Jun 2002.

[61] Gavin David Scott and Douglas Natelson. Kondo resonances in molecular devices. *ACS Nano*, 4(7):3560–3579, 2010.

[62] Riccardo Frisenda, Rocco Gaudenzi, Carlos Franco, Marta Mas-Torrent, Concepció Rovira, Jaume Veciana, Isaac Alcon, Stefan T. Bromley, Enrique Burzurí, and Herre S. J. van der Zant. Kondo effect in a neutral and stable all organic radical single molecule break junction. *Nano Letters*, 15(5):3109–3114, 2015.

[63] E. Schrödinger. Die gegenwärtige situation in der quantenmechanik. *Naturwissenschaften*, 23(48):807–812, 1935.

[64] David J. Wineland. Nobel lecture: Superposition, entanglement, and raising schrödinger's cat. *Rev. Mod. Phys.*, 85:1103–1114, Jul 2013.

[65] Sandrine Sans-Lenain and Alain Gleizes. Structural features of homo- and heteroleptic complexes of copper(II) with 2,2,6,6-tetramethyl-3,5-heptanedione and 3-chloro-2,4-pentanedione. *Inorganica Chimica Acta*, 211(1):67–75, 1993.

[66] Christian Joachim, James K. Gimzewski, Reto R. Schlittler, and Corinne Chavy. Electronic transparence of a single C_{60} molecule. *Phys. Rev. Lett.*, 74:2102–2105, Mar 1995.

[67] A. F. Takács, F. Witt, S. Schmaus, T. Balashov, M. Bowen, E. Beaurepaire, and W. Wulfhekel. Electron transport through single phthalocyanine molecules studied using scanning tunneling microscopy. *Phys. Rev. B*, 78:233404, Dec 2008.

[68] R. Temirov, A. Lassise, F. B. Anders, and F. S. Tautz. Kondo effect by controlled cleavage of a single-molecule contact. *Nanotechnology*, 19:065401, 2008.

[69] Stefan Schmaus, Alexei Bagrets, Yasmine Nahas, Toyo K Yamada, Annika Bork, Martin Bowen, Eric Beaurepaire, Ferdinand Evers, and Wulf Wulfhekel. Giant magnetoresistance through a single molecule. *Nature Nanotechnology*, 6(3):185–189, 2011.

[70] Sujoy Karan, David Jacob, Michael Karolak, Christian Hamann, Yongfeng Wang, Alexander Weismann, Alexander I. Lichtenstein, and Richard Berndt. Shifting the voltage drop in electron transport through a single molecule. *Phys. Rev. Lett.*, 115:016802, Jul 2015.

[71] Bingqian Xu, Xiaoyin Xiao, and Nongjian J. Tao. Measurements of single-molecule electromechanical properties. *Journal of the American Chemical Society*, 125(52):16164–16165, 2003.

[72] Youngsang Kim, Hyunwook Song, Florian Strigl, Hans-Fridtjof Pernau, Takhee Lee, and Elke Scheer. Conductance and vibrational states of single-molecule junctions controlled by mechanical stretching and material variation. *Phys. Rev. Lett.*, 106:196804, May 2011.

[73] Makusu Tsutsui, Takanori Morikawa, Yuhui He, Akihide Arima, and Masateru Taniguchi. High thermopower of mechanically stretched single-molecule junctions. *Scientific Reports*, 5:11519, Jun 2015.

[74] Markus Ternes. Spin excitations and correlations in scanning tunneling spectroscopy. *New Journal of Physics*, 17(6):063016, 2015.

[75] F. D. M. Haldane. Scaling theory of the asymmetric Anderson model. *Phys. Rev. Lett.*, 40:416–419, Feb 1978.

[76] P. Wahl, L. Diekhöner, M. A. Schneider, L. Vitali, G. Wittich, and K. Kern. Kondo temperature of magnetic impurities at surfaces. *Phys. Rev. Lett.*, 93:176603, Oct 2004.

[77] J. Fernández-Rossier. Theory of single-spin inelastic tunneling spectroscopy. *Phys. Rev. Lett.*, 102:256802, Jun 2009.

[78] E. del Barco, A. D. Kent, E. C. Yang, and D. N. Hendrickson. Quantum superposition of high spin states in the single molecule magnet Ni_4. *Phys. Rev. Lett.*, 93:157202, Oct 2004.

[79] C. Monroe, D. M. Meekhof, B. E. King, and D. J. Wineland. A "schrödinger cat" superposition state of an atom. *Science*, 272(5265):1131–1136, 1996.

[80] Magnus Paulsson, Casper Krag, Thomas Frederiksen, and Mads Brandbyge. Conductance of alkanedithiol single-molecule junctions: A molecular dynamics study. *Nano Letters*, 9(1):117–121, 2009.

[81] Guillem Aromí, David Aguilà, Patrick Gamez, Fernando Luis, and Olivier Roubeau. Design of magnetic coordination complexes for quantum computing. *Chem. Soc. Rev.*, 41:537–546, 2012.

[82] A. Braun and J. Tcherniac. über die Produkte der Einwirkung von Acetanhydrid auf Phthalamid. *Berichte der deutschen chemischen Gesellschaft*, 40(2):2709–2714, 1907.

[83] N.B. McKeown. *Phthalocyanine Materials: Synthesis, Structure and Function*. Bruce Dunn. Cambridge University Press, 1998.

[84] G. Guillaud, J. Simon, and J.P. Germain. Metallophthalocyanines: Gas sensors, resistors and field effect transistors. *Coordination Chemistry Reviews*, 178-180(Part 2):1433–1484, 1998.

[85] Arrigo Calzolari, Andrea Ferretti, and Marco Buongiorno Nardelli. Ab initio correlation effects on the electronic and transport properties of metal(II)-phthalocyanine-based devices. *Nanotechnology*, 18(42):424013, 2007.

[86] P. H. Lippel, R. J. Wilson, M. D. Miller, Ch. Wöll, and S. Chiang. High-resolution imaging of copper-phthalocyanine by scanning-tunneling microscopy. *Phys. Rev. Lett.*, 62:171–174, Jan 1989.

[87] J. Michael Gottfried. Surface chemistry of porphyrins and phthalocyanines. *Surface Science Reports*, 70(3):259–379, 2015.

[88] Aidi Zhao, Qunxiang Li, Lan Chen, Hongjun Xiang, Weihua Wang, Shuan Pan, Bing Wang, Xudong Xiao, Jinlong Yang, J. G. Hou, and Qingshi Zhu. Controlling the Kondo effect of an adsorbed magnetic ion through its chemical bonding. *Science*, 309(5740):1542–1544, 2005.

[89] Ying-Shuang Fu, Shuai-Hua Ji, Xi Chen, Xu-Cun Ma, Rui Wu, Chen-Chen Wang, Wen-Hui Duan, Xiao-Hui Qiu, Bo Sun, Ping Zhang, Jin-Feng Jia, and Qi-Kun Xue. Manipulating the Kondo resonance through quantum size effects. *Phys. Rev. Lett.*, 99:256601, Dec 2007.

[90] Hiroyuki Yamada, Toshihiro Shimada, and Atsushi Koma. Preparation and magnetic properties of manganese(II) phthalocyanine thin films. *The Journal of Chemical Physics*, 108(24):10256–10261, 1998.

[91] S. Heutz, C. Mitra, W. Wu, A. J. Fisher, A. Kerridge, M. Stoneham, A. H. Harker, J. Gardener, H.-H. Tseng, T. S. Jones, C. Renner, and G. Aeppli. Molecular thin films: A new type of magnetic switch. *Advanced Materials*, 19(21):3618–3622, 2007.

[92] Wei Wu, N. M. Harrison, and A. J. Fisher. Electronic structure and exchange interactions in cobalt-phthalocyanine chains. *Phys. Rev. B*, 88:024426, Jul 2013.

[93] Manuel Gruber, Fatima Ibrahim, Samy Boukari, Hironari Isshiki, Loïc Joly, Moritz Peter, Michał Studniarek, Victor Da Costa, Hashim Jabbar, Vincent Davesne, et al. Exchange bias and room-temperature magnetic order in molecular layers. *Nature Materials*, 14:981, Jul 2015.

[94] Haiming Guo, Yeliang Wang, Shixuan Du, and Hong jun Gao. High-resolution scanning tunneling microscopy imaging of si(1 1 1)-7$\times$ 7 structure and intrinsic molecular states. *Journal of Physics: Condensed Matter*, 26(39):394001, 2014.

[95] Yan Feng Zhang, Hironari Isshiki, Keiichi Katoh, Yusuke Yoshida, Masahiro Yamashita, Hitoshi Miyasaka, Brian K. Breedlove, Takashi Kajiwara, Shinya Takaishi, and Tadahiro Komeda. Low-temperature scanning tunneling microscopy investigation of bis(phthalocyaninato)yttrium growth on Au(111): From individual molecules to two-dimensional domains. *The Journal of Physical Chemistry C*, 113(22):9826–9830, 2009.

[96] Naoto Ishikawa, Miki Sugita, Tadahiko Ishikawa, Shin-ya Koshihara, and Youkoh Kaizu. Lanthanide Double-Decker Complexes Functioning as Magnets at the Single-Molecular Level. *Journal of the American Chemical Society*, 125(29):8694–8695, 2003.

[97] Carolina R. Ganivet, Beatriz Ballesteros, Gema de la Torre, Juan M. Clemente-Juan, Eugenio Coronado, and Tomás Torres. Influence of peripheral substitution on the

magnetic behavior of single-ion magnets based on homo- and heteroleptic Tb^{III} bis(phthalocyaninate). *Chemistry - A European Journal*, 19(4):1457–1465, 2013.

[98] Jeffrey D. Rinehart, Ming Fang, William J. Evans, and Jeffrey R. Long. Strong exchange and magnetic blocking in N_2^{3-}-radical-bridged lanthanide complexes. *Nature Chemistry*, 3:538, May 2011.

[99] Constantinos J. Milios, Alina Vinslava, Wolfgang Wernsdorfer, Stephen Moggach, Simon Parsons, Spyros P. Perlepes, George Christou, and Euan K. Brechin. A record anisotropy barrier for a single-molecule magnet. *Journal of the American Chemical Society*, 129(10):2754–2755, 2007.

[100] Humphrey L.C. Feltham and Sally Brooker. Review of purely $4f$ and mixed-metal nd-$4f$ single-molecule magnets containing only one lanthanide ion. *Coordination Chemistry Reviews*, 276(Supplement C):1–33, 2014.

[101] A. J. Heinrich, J. A. Gupta, C. P. Lutz, and D. M. Eigler. Single-Atom Spin-Flip Spectroscopy. *Science*, 306(5695):466–469, 2004.

[102] Cyrus F. Hirjibehedin, Chiung-Yuan Lin, Alexander F. Otte, Markus Ternes, Christopher P. Lutz, Barbara A. Jones, and Andreas J. Heinrich. Large Magnetic Anisotropy of a Single Atomic Spin Embedded in a Surface Molecular Network. *Science*, 317(5842):1199–1203, 2007.

[103] Focko Meier, Lihui Zhou, Jens Wiebe, and Roland Wiesendanger. Revealing magnetic interactions from single-atom magnetization curves. *Science*, 320(5872):82–86, 2008.

[104] Tadahiro Komeda, Hironari Isshiki, Jie Liu, Yan-Feng Zhang, Nicolás Lorente, Keiichi Katoh, Brian K. Breedlove, and Masahiro Yamashita. Observation and electric current control of a local spin in a single-molecule magnet. *Nature Communications*, 2:217, Mar 2011.

[105] Sarah Fahrendorf, Nicolae Atodiresei, Claire Besson, Vasile Caciuc, Frank Matthes, Stefan Blügel, Paul Kögerler, Daniel E. Bürgler, and Claus M. Schneider. Accessing $4f$-states in single-molecule spintronics. *Nature Communications*, 4:2425, Sep 2013.

[106] Ben Warner, Fadi El Hallak, Nicolae Atodiresei, Philipp Seibt, Henning Prüser, Vasile Caciuc, Michael Waters, Andrew J Fisher, Stefan Blügel, Joris Van Slageren, et al. Sub-molecular modulation of a $4f$ driven Kondo resonance by surface-induced asymmetry. *Nature Communications*, 7:12785, Sep 2016.

[107] Lucia Vitali, Stefano Fabris, Adriano Mosca Conte, Susan Brink, Mario Ruben, Stefano Baroni, and Klaus Kern. Electronic structure of surface-supported bis(phthalocyaninato) terbium(III) single molecular magnets. *Nano Letters*, 8(10):3364–3368, 2008.

[108] Hironari Isshiki, Jinjie Chen, Kevin Edelmann, and Wulf Wulfhekel. Sub-monolayer film growth of a volatile lanthanide complex on metallic surfaces. *Beilstein Journal of Nanotechnology*, 6:2412–2416, 2015.

[109] Wen Yu, Frank Schramm, Eufemio Moreno Pineda, Yanhua Lan, Olaf Fuhr, Jinjie Chen, Hironari Isshiki, Wolfgang Wernsdorfer, Wulf Wulfhekel, and Mario Ruben. Single-molecule magnet behavior in 2,2'-bipyrimidine-bridged dilanthanide complexes. *Beilstein Journal of Nanotechnology*, 7:126–137, Jan 2016.

[110] Mark J. Hampden-Smith and Toivo T. Kodas. Chemical vapor deposition of metals: Part 1. An overview of CVD processes. *Chemical Vapor Deposition*, 1(1):8–23, 1995.

[111] Sabrina Munery, Nicolas Ratel-Ramond, Youness Benjalal, Loranne Vernisse, Olivier Guillermet, Xavier Bouju, Roland Coratger, and Jacques Bonvoisin. Synthesis and characterization of a series of ruthenium tris(β-diketonato) complexes by an UHV-STM investigation and numerical calculations. *European Journal of Inorganic Chemistry*, 2011(17):2698–2705, 2011.

[112] Juan E. Sicre, J. T. Dubois, K. J. Eisentraut, and R. E. Sievers. Volatile lanthanide chelates. II. Vapor pressures, heats of vaporization, and heats of sublimation. *Journal of the American Chemical Society*, 91(13):3476–3481, 1969.

[113] Robert E. Sievers and Jean E. Sadlowski. Volatile Metal Complexes. *Science*, 201(4352):217–223, 1978.

[114] Valery V. Sliznev, Natalya V. Belova, and Georgiy V. Girichev. The molecular structure of tris(dipivaloylmethanato) lanthanides, $Ln(thd)_3$ (Ln=La-Lu): Peculiarities and regularities. *Computational and Theoretical Chemistry*, 1055:78–87, 2015.

[115] Nicholas F. Chilton, Stuart K. Langley, Boujemaa Moubaraki, Alessandro Soncini, Stuart R. Batten, and Keith S. Murray. Single molecule magnetism in a family of mononuclear β-diketonate lanthanide(iii) complexes: rationalization of magnetic anisotropy in complexes of low symmetry. *Chem. Sci.*, 4:1719–1730, 2013.

[116] Shobhana Narasimhan and David Vanderbilt. Elastic stress domains and the herringbone reconstruction on Au(111). *Phys. Rev. Lett.*, 69:1564–1567, Sep 1992.

[117] Yokoyama, Takashi, Yokoyama, Shiyoshi, Kamikado, Toshiya, Okuno, Yoshishige, and Mashiko, Shinro. Selective assembly on a surface of supramolecular aggregates with controlled size and shape. *Nature*, 413(6856):619–621, Oct 2001.

[118] Matthias Böhringer, Karina Morgenstern, Wolf-Dieter Schneider, Richard Berndt, Francesco Mauri, Alessandro De Vita, and Roberto Car. Two-dimensional self-assembly of supramolecular clusters and chains. *Phys. Rev. Lett.*, 83:324–327, Jul 1999.

[119] L. C. Davis, M. P. Everson, R. C. Jaklevic, and Weidian Shen. Theory of the local density of surface states on a metal: Comparison with scanning tunneling spectroscopy of a Au(111) surface. *Phys. Rev. B*, 43:3821–3830, Feb 1991.

[120] M. F. Crommie, C. P. Lutz, and D. M. Eigler. Imaging standing waves in a two-dimensional electron gas. *Nature*, 363(6429):524–527, Jun 1993.

[121] Ferdous Ara, Zhi Kun Qi, Jie Hou, Tadahiro Komeda, Keiichi Katoh, and Masahiro Yamashita. A scanning tunneling microscopy study of the electronic and spin states of bis(phthalocyaninato)terbium(III) ($TbPc_2$) molecules on Ag(111). *Dalton Trans.*, 45:16644–16652, 2016.

[122] David Aguilà, Leoní A. Barrios, Verónica Velasco, Olivier Roubeau, Ana Repollés, Pablo J. Alonso, Javier Sesé, Simon J. Teat, Fernando Luis, and Guillem Aromí. Heterodimetallic [LnLn'] lanthanide complexes: Toward a chemical design of two-qubit molecular spin quantum gates. *Journal of the American Chemical Society*, 136(40):14215–14222, 2014.

[123] Stefan Wagner, Ferdinand Kisslinger, Stefan Ballmann, Frank Schramm, Rajadurai Chandrasekar, Tilmann Bodenstein, Olaf Fuhr, Daniel Secker, Karin Fink, Mario Ruben, and Heiko B. Weber. Switching of a coupled spin pair in a single-molecule junction. *Nature Nanotechnology*, 8:575, Jul 2013.

[124] L. Zhang, A. Bagrets, D. Xenioti, R. Korytár, M. Schackert, T. Miyamachi, F. Schramm, O. Fuhr, R. Chandrasekar, M. Alouani, M. Ruben, W. Wulfhekel, and F. Evers. Kondo effect in binuclear metal-organic complexes with weakly interacting spins. *Phys. Rev. B*, 91:195424, May 2015.

[125] Manuel Gruber, Toshio Miyamachi, Vincent Davesne, Martin Bowen, Samy Boukari, Wulf Wulfhekel, Mebarek Alouani, and Eric Beaurepaire. Spin crossover in $Fe(phen)_2(NCS)_2$ complexes on metallic surfaces. *The Journal of Chemical Physics*, 146(9):092312, 2017.

[126] Saber Gueddida, Manuel Gruber, Toshio Miyamachi, Eric Beaurepaire, Wulf Wulfhekel, and Mebarek Alouani. Exchange coupling of spin-crossover molecules to ferromagnetic co islands. *The Journal of Physical Chemistry Letters*, 7(5):900–904, 2016.

[127] W. A. Baker and H. M. Bobonich. Magnetic properties of some high-spin complexes of iron(II). *Inorganic Chemistry*, 3(8):1184–1188, 1964.

[128] E. König and K. Madeja. Unusual magnetic behaviour of some iron(II)-bis-(1,10-phenanthroline) complexes. *Chem. Commun. (London)*, pages 61–62, 1966.

[129] V. Davesne, M. Gruber, T. Miyamachi, V. Da Costa, S. Boukari, F. Scheurer, L. Joly, P. Ohresser, E. Otero, F. Choueikani, A. B. Gaspar, J. A. Real, W. Wulfhekel, M. Bowen, and E. Beaurepaire. First glimpse of the soft x-ray induced excited spin-state trapping effect dynamics on spin cross-over molecules. *The Journal of Chemical Physics*, 139(7):074708, 2013.

[130] Bernard Gallois, Jose Antonio Real, Christian Hauw, and Jacqueline Zarembowitch. Structural changes associated with the spin transition in bis(isothiocyanato)bis(1,10-phenanthroline)iron: a single-crystal x-ray investigation. *Inorganic Chemistry*, 29(6):1152–1158, 1990.

[131] Manuel Gruber, Vincent Davesne, Martin Bowen, Samy Boukari, Eric Beaurepaire, Wulf Wulfhekel, and Toshio Miyamachi. Spin state of spin-crossover complexes: From single molecules to ultrathin films. *Phys. Rev. B*, 89:195415, May 2014.

[132] Y. Takahashi, T. Miyamachi, K. Ienaga, N. Kawamura, A. Ernst, and F. Komori. Orbital selectivity in scanning tunneling microscopy: Distance-dependent tunneling process observed in iron nitride. *Phys. Rev. Lett.*, 116:056802, Feb 2016.

[133] Y. Takahashi, T. Miyamachi, S. Nakashima, N. Kawamura, Y. Takagi, M. Uozumi, V. N. Antonov, T. Yokoyama, A. Ernst, and F. Komori. Thickness-dependent electronic and magnetic properties of $\gamma'-Fe_4N$ atomic layers on Cu(001). *Phys. Rev. B*, 95:224417, Jun 2017.

[134] G.H. Jonker and J.H. Van Santen. Ferromagnetic compounds of manganese with perovskite structure. *Physica*, 16(3):337–349, 1950.

[135] A. M. Haghiri-Gosnet, J. Wolfman, B. Mercey, Ch. Simon, P. Lecoeur, M. Korzenski, M. Hervieu, R. Desfeux, and G. Baldinozzi. Microstructure and magnetic properties of strained $La_{0.7}Sr_{0.3}MnO_3$ thin films. *Journal of Applied Physics*, 88(7):4257–4264, 2000.

Acknowledgement

I would like to deliver my acknowledgements to all of those who helped and enabled me to finish this dissertation.

First and foremost, my respectful gratitude goes to my supervisor, Prof. Wulf Wulfhekel for offering me the opportunity to perform these striking projects. His valuable guidance and constant encouragement about not only science but also life lead me towards the proper direction in years.

Secondly, I shall extend my heartfelt thanks to Dr. Hironari Isshiki, who led me into the world of studying molecules with STM, for all his unwavering help and scholarly advice.

I feel so grateful to Timo Frauhammer, Max Suppes, Ryohei Nemoto for the efficient and effective teamwork on common experiments. Especially to Timo Frauhammer, who is carrying on the succeeding projects of this dissertation.

I sincerely thank my collaborators on different projects. They are Prof. Mario Ruben, Prof. Wolfgang Wernsdorfer, Dr. Wen Yu, and Dr. Frank Schramm from Karlsruhe; Prof. Eric Beaurepaire, Dr. Martin Bowen, and Dr. Manuel Gruber from Strasbourg; Dr. Toshio Miyamachi from Tokyo; Prof. Alek Dediu and Dr. Patrizio Graziosi from Bologna. Without these collaborations, I could not make this dissertation a reality.

Special thanks should go to my second referee, Prof. Wolfgang Wenzel. Without his report, I could not complete this dissertation in time.

At the same time, high tribute should be paid to Timofey, Timo, Fang, Vibhuti, Nikhil, and David for proofreading this dissertation.

I do owe much of my sincere gratitude to my lovely current and former group members, Timofey, Lukas, Marie, Hironari, Fang, Moritz, Tobias Märkl, Bastian, Jasmin, Kevin, Rafael, Timo, Sergey, Vibhuti, Martin, Jonas, Clemens, Tobias Engelhardt, Dominik, Lars, Loïc, Endre, Julian Vincent, as well as my institute colleagues. All of them gave me their best help and contributed to the homelike atmosphere spreading from the group to the whole institute. I really enjoyed the stay in the AG Wulfhekel, in Physikalischens Institut, in KIT, in Germany.

Last but not least, my thanks should also go to my beloved family and friends, particularly my girlfriend Xiaomeng, for their loving support and invariable encouragement all through the four years.

Experimental Condensed Matter Physics
(ISSN 2191-9925)

Herausgeber
Physikalisches Institut

Prof. Dr. David Hunger
Prof. Dr. Alexey Ustinov
Prof. Dr. Georg Weiß
Prof. Dr. Wolfgang Wernsdorfer
Prof. Dr. Wulf Wulfhekel

Band 1 Alexey Feofanov
Experiments on flux qubits with pi-shifters. 2011
ISBN 978-3-86644-644-1

Band 2 Stefan Schmaus
Spintronics with individual metal-organic molecules. 2011
ISBN 978-3-86644-649-6

Band 3 Marc Müller
Elektrischer Leitwert von magnetostriktiven Dy-Nanokontakten. 2011
ISBN 978-3-86644-726-4

Band 4 Torben Peichl
Einfluss mechanischer Deformation auf atomare Tunnelsysteme – untersucht mit Josephson Phasen-Qubits. 2012
ISBN 978-3-86644-837-7

Band 5 Dominik Stöffler
Herstellung dünner metallischer Brücken durch Elektromigration und Charakterisierung mit Rastersondentechniken. 2012
ISBN 978-3-86644-843-8

Band 6 Tihomir Tomanic
Untersuchung des elektronischen Oberflächenzustands von Ag-Inseln auf supraleitendem Niob (110). 2012
ISBN 978-3-86644-898-8

Band 7 Lukas Gerhard
Magnetoelectric coupling at metal surfaces. 2013
ISBN 978-3-7315-0063-6

Band 8 Kirill Fedorov
Fluxon readout for superconducting flux qubits. (nicht erschienen)
ISBN 978-3-7315-0067-4

Die Bände sind unter www.ksp.kit.edu als PDF frei verfügbar oder als Druckausgabe bestellbar.

Band 9 Jochen Zimmer
Cooper pair transport in arrays of Josephson junctions. 2014
ISBN 978-3-7315-0130-5

Band 10 Oliver Berg
Elektrischer Transport durch Nanokontakte von Selten-Erd-Metallen. 2014
ISBN 978-3-7315-0209-8

Band 11 Grigorij Jur'evic Grabovskij
Investigation of coherent microscopic defects inside the tunneling barrier of a Josephson junction. 2014
ISBN 978-3-7315-0210-4

Band 12 Cornelius Thiele
STM Characterization of Phenylene-Ethynylene Oligomers on Au(111) and their Integration into Carbon Nanotube Nanogaps. 2014
ISBN 978-3-7315-0235-7

Band 13 Michael Peter Schackert
Scanning Tunneling Spectroscopy on Electron-Boson Interactions in Superconductors. 2014
ISBN 978-3-7315-0238-8

Band 14 Susanne Butz
One-Dimensional Tunable Josephson Metamaterials. 2014
ISBN 978-3-7315-0271-5

Band 15 Philipp Jung
Nonlinear Effects in Superconducting Quantum Interference Meta-Atoms. 2014
ISBN 978-3-7315-0294-4

Band 16 Sebastian Probst
Hybrid quantum system based on rare earth doped crystals. 2016
ISBN 978-3-7315-0345-3

Band 17 Wolfram Kittler
Magnetische Anisotropie und Quantenphasenübergang in $CeTi_{1-x}V_xGe_3$. 2015
ISBN 978-3-7315-0363-7

Band 18 Moritz Peter
Towards magnetic resonance in scanning tunneling microscopy using heterodyne detection. 2015
ISBN 978-3-7315-0410-8

Band 19 Junji Tobias Märkl
Investigation of Magnetic Adatoms with Scanning Tunneling Techniques. 2015
ISBN 978-3-7315-0435-1

Band 20 Jochen Braumüller
Quantum simulation experiments with superconducting circuits. 2018
ISBN 978-3-7315-0780-2

Band 21 Jinjie Chen
Local Investigation of Single Magnetic Molecules with Scanning Tunneling Microscopy. 2018
ISBN 978-3-7315-0819-9
